New Cambrid
MATHEMATI

■ ▲ ● **MODULE 1** ● ▲ ■

Games and mats
Teacher's handbook

■

Sue Atkinson

▲

Wendy Garrard

●

Lynne McClure

■

Sharon Harrison

▲

CAMBRIDGE
UNIVERSITY PRESS

Published by the Press Syndicate of the University of Cambridge
The Pitt Building, Trumpington street, Cambridge CB2 1RP
40 West 20th Street, New York, NY 10011 – 4211, USA
10 Stamford Road, Oakleigh, Melbourne 3166, Australia

First published 1995

Printed in Great Britain by GreenShires Print Ltd, Kettering

A catalogue record for this book is available from the British Library

ISBN 0 521 47190 7

Notice to teachers

Introduction to the games

The games are designed to provide consolidation of mathematical ideas in an enjoyable way. They can also be used to encourage co-operative working and to develop powers of reasoning by thinking about strategies. References to games relevant to particular activities are given in the Teacher's resource book.

How to prepare the games

The following grid shows the base boards, sheets of cut-ups and equipment needed for each game. A plastic bag and label is provided for each one. It is a good idea to include a copy of the rules in the bag as well as the boards, cut-ups and equipment. The cut-ups should be cut along the *dashed* lines only to make cards.

How to use the games

The games can be used in the classroom to provide further practice or they can be sent home for children to play with their families. It is a good idea to play the games with the children in order to introduce the rules. Each board can be used for several different games and suggestions are provided which generally increase in difficulty. Of course, the children can also make up their own rules.

Children's own games

Page 3 can be photocopied to give children the opportunity to make up their own mathematical track game. You might want to suggest a context, such as a game of racing cars or bikes, or something connected with your class topic. Wrapping paper pictures or pictures from greetings cards can be stuck onto a copy of the sheet. Encourage children to make up their own rules.

Number cards

Among the cut-ups are two sets of 1 to 24 number cards. These can be used to support the activities in the Teacher's resource book and also for games made up by you or the children.

1 ■ ● ▲ ◀

Equipment given in italics is not supplied in the pack.
Some non standard dice are required. Blank dice and stickers are
supplied, but you may prefer to write on a larger wooden cube.

Title	Base boards	Cut-ups	Equipment	Mathematical content
Dragon and friends	4	4 different	*Pencil and paper* for Game 5 *A calculator* if wanted for Game 5	matching, sorting, counting
Monster footprints	1	1	10 green, 10 yellow counters 1 blank dice (labelled 0, 1, 1, 2, 2, jump) 1 dice (1–6) *12 unifix or other small objects* for Game 4	counting on, counting
Spotty monster	1		25 red, 25 blue, 4 yellow, 4 green counters 1 blank dice (labelled 0, 1, 1, 1, 2, 2) 2 dice (1–6) *A calculator* if wanted for Game 3	counting to 6, subtraction
Cat and mouse	1	1	1 green, 1 yellow counters	2D shapes
Three bears	1	1	9 red, 9 blue counters *12 unifix or other small objects* for Game 5	matching
Monster's quilt	1		10 yellow, 10 green, 1 red, 1 blue counters 1 blank dice (labelled 3, 3, 4, 4, 5, 6)	2D shapes
Bugs	1	1	1 ten-sided dice (0–9)	counting
Necklaces	1		12 red, 12 yellow, 12 blue, 12 green counters 1 blank dice (labelled 0, 0, 1, 1, 2, 2) 1 blank dice (labelled 1, 1, 2, 2, 3, 3) 1 dice (1–6)	counting, number bonds to 6
Race to the honey	1		1 red, 1 yellow, 1 blue, 1 green counters 1 dice (1–6)	counting on, numbers to 100
Shopping	2	2 different	1 dice (1–6) 1 blank dice (labelled 0, 1, 1, 2, 2, 2) *25 one pence , 10 ten pence coins*	money
Dominoes		2½ different		recognising numerals and number words, counting
What's the time, Mr Wolf?	1	1½ different	1 red, 1 yellow, 1 blue, 1 green counters	telling the time
Snakes	1		10 yellow, 10 green counters 1 blank dice (labelled 5, 6, 7, 8, 9, 10) 1 blank dice (labelled 0, 1, 2, 3, 4, 5) *A copy of page 19 for each player* for Game 5 *20 unifix or other small objects* for Game 4 *Crayons* for Game 5	counting, adding to 10

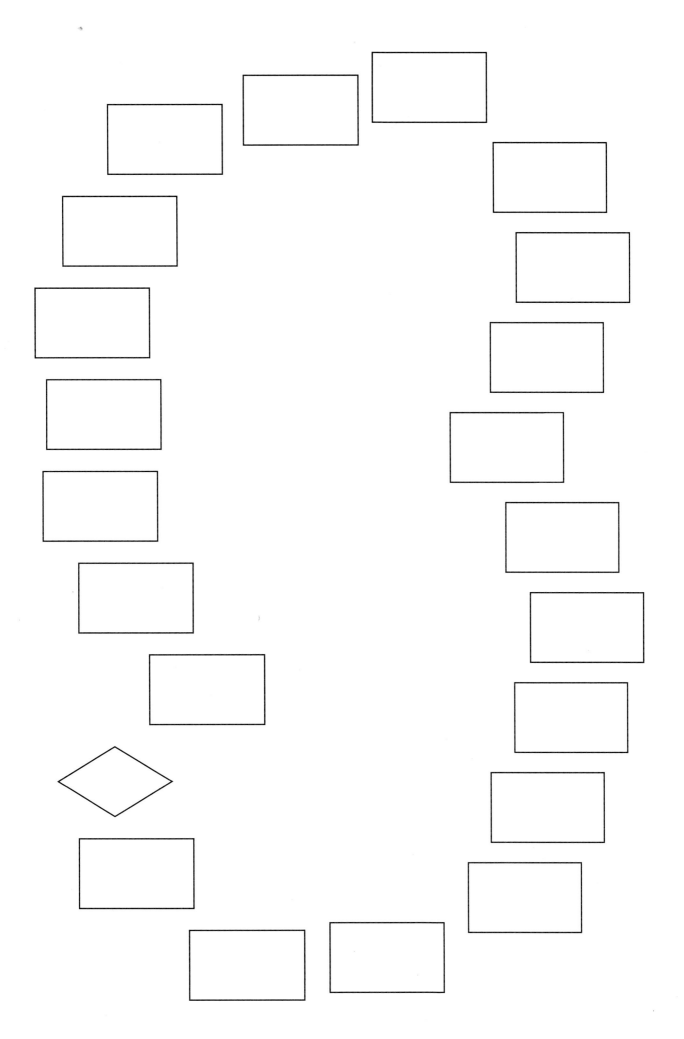

Dragon and friends

Game 1 ■ Matching

You need: 2 to 4 players
a game board for each player
the corresponding sets of cards, all shuffled
together

Put the cards in a pile face down on the table.

Take a card in turns. If you have that picture on
your board, put the card on top of it.

If you don't have that picture, put the card on a
second pile.

If you reach the end of the pile of cards, shuffle the
discard pile and start again.

The winner is the first to match all their pictures.

Game 2 ▲ Secret friends

You need: 2 to 8 players
4 game boards
all the cards, shuffled together

One player has all the game boards and secretly
chooses just one of the friends (for example, little
ted with shorts and a hat and an ice cream).

The cards are dealt out to the other players who put
them face up on the table.

These other players then take it in turns to try to
guess the secret friend. Ask questions that need the
reply 'yes' or 'no' only, for example:

Is it a monster?

Is it a teddy bear?

Does it have an ice cream?

If you think you know who the friend is (even if
you don't have the card), you can make a guess
when it is your turn.

Game 3 ● Dominoes

You need: 2 to 4 players
all the cards, shuffled together

Play dominoes with the cards. One card matches
onto another if it has something similar, so monster
with a hat could match onto dragon with a hat or a
monster with a balloon.

Deal out all the cards and take it in turns to place
one at either end of the line. You must say what it is
that you are matching.

The winner is the first ꞏ ꞏ ꞏ +down all their
cards, or the player left with ꞏꞏ ꞏwest cards at the
end.

Game 4 ■ A sorting game

You need: 1 or 2 players
some or all of the cards

One player or two players can work together and
sort the friends into groups.

Decide beforehand how many groups you want, for
example, you might decide to make two groups and
place together all the friends with a balloon and all
the friends without a balloon.

Some ways of sorting will leave you with cards that
won't 'go'. If you choose 'all the friends ready for
the beach' and 'all the friends going out to play in
the rain' you might have some that won't fit into
the groups.

Game 5 ▲ A counting game

You need: 1 to 8 players
1 game board or a corresponding set of
cards for each player
pencil and paper for each player
a calculator, if required

Choose one, two or three things to count, for
example, all the balloons, all the ice creams and all
the boots.

Count those things on your board or set of cards.
Two players could count the same things. Use a
calculator if you want.

Find a way to put your count onto paper so that
you won't forget the total.

Players can check if others are right.

The paper can be put on a board, or in a book to
keep for the next time someone counts those same
things. Do they agree?

Children could progress to counting objects on all
the boards.

Game 6 ● Snap

You need: 2 to 4 players
all the cards, shuffled together

Deal out all the cards to the players.

Decide what will make a 'snap'. For example, any
'snap' might count provided you can say why, or a
simple version would be a snap if it is the same
kind of friend, for example, another dragon.

Take it in turns to show one card at a time and put
it face up on a pile in front of you.

If it matches the last card played the first player to
call 'snap' wins all the cards in the two piles.

The winner is the player with the most cards at the
end.

Monster footprints

■ Game 1

You need: 2 players
the game board
1 green, 1 yellow counter
a dice labelled 0, 1, 1, 2, 2, jump

Choose a track each and put a counter on your start.

Take turns to throw the dice.

If you throw a number, move your counter that number of spaces.

If the dice shows 'jump', jump your counter across the river onto the corresponding number on the other track. If that space already has a counter, send it back to the start. (You may want to ignore this rule for very young children as it can be rather disheartening.)

The winner is the first player to land on their *own* ice cream.

▲ Game 2

You need: 2 players
the game board
the 1–10 and ice-cream cards, shuffled
1 green, 1 yellow counter

Choose a track each and put a counter on your start.

Place the cards in a pile face down.

Take it in turns to turn over a card.

You may move your counter one step at a time, but only when the next number is turned over. For example, if your counter is on 5 then you may only move when you turn up a 6.

The winner is the first player to reach the cone.

● Game 3

You need: 2 players
the game board
the 1–10 cards, shuffled
10 green, 10 yellow counters

Cover both the tracks with counters so that the numbers are hidden. Use a different colour for each player.

Place the cards in a pile face down. Take it in turns to turn up a card.

Remove the counter covering the number turned up. (You can decide a rule for incorrect removal of counters.)

The winner is the first player to remove all their counters.

■ Game 4 (Extension)

You need: 2 players
the game board
1 green, 1 yellow counter
1 dice (1–6)
12 unifix or other small objects

Choose a track each and put a counter on your start.

Take turns to throw the dice and pick up that number of cubes.

Move one space each time a group of 4 cubes (4 toes) can be made. After the move, return 4 cubes to the pile.

The winner is the first player to reach footprint number 10.

Spotty monster

■ Game 1

You need: 2 players
the game board
10 red, 10 blue counters
a dice labelled 0, 1, 1, 1, 2, 2

Each player puts a counter on each spot of an arm (10 counters).

Throw the dice and remove that many counters.

The winner is the first player to remove all their counters.

▲ Game 2

You need: 2 players
the game board
1 red, 1 blue counter
1 dice (1–6)

Each choose one of monster's hands and put your counter on it.

Take turns to throw the dice and race to reach monster's toe. If your counter lands on a red spot on monster's arm, miss a go.

● Game 3 (Extension)

You need: 2 players
the game board
up to 25 red, 25 blue counters
2 dice (1–6)
a calculator

This is played like game 1. Start by putting your colour counters on the monster. You can put them just on one arm, or put up to 25 counters somewhere on your side of the monster.

Throw both the dice and find the difference between the two numbers. Take off that many counters. So if the dice show 6 and 2, the difference is 4 (6 – 2 = 4) and you can take off 4 of your counters. A calculator can be used to help.

The winner is the first player to take off all their counters.

■ Game 4 (Extension)

You need: 2 players
the game board
up to 25 red, 25 blue counters
2 dice (1–6)

This game is played just like game 3, but if your difference is 4, you can choose to

take off 4 of your counters

or

add 4 of your opponent's counters to the board

or

split up 4 and take off perhaps 2 of yours and add on 2 of your opponent's counters, provided they do not end up with more than 25 of their counters on the board.

The first player to remove all their counters is the winner.

▲ Game 5 (Extension)

You need: 2, 3 or 4 players
the game board
2, 3 or 4 counters for each player
(a different colour for each player)
1 dice (1–6)

Each player puts 4 (or 3 or 2) counters on one of the toys monster has in her hands.

Take turns to throw the dice. If you throw 4, say, you move one of your counters 4 spaces, or you could choose to move 2 counters and move one 3 spaces and the other 1 space. You may not have more than one counter on a spot.

If you want to move onto a spot that someone else is already on, you can remove that counter and put it on the sky. That counter is now out of the game.

The object is to race all your counters down to either foot and end up with as many as you can on the grass.

The player with the most counters on the grass is the winner. If there is a tie the player to have their counters home first is the winner.

● Game 6

Invent your own game.

Cat and mouse

■ Game 1

You need: 2 players
the game board
all the cards, shuffled together
1 green, 1 yellow counter

Put the cards in a pile face down.

One person is the cat and one is the mouse. Start in your home corner (top left for the mouse and bottom right for the cat).

Take it in turns to take the card at the top of the pack. If you pick a card that has a cheese on it you must move one space, up, down, left, right or diagonally to a cheese with that same shape, if you can.

If you pick a card that has all five cheeses on it, you can move to any adjacent cheese.

The used cards are made into another pile. These are shuffled when all the first pile has been used.

The aim for the mouse is to get to the cat's home to get the best cheese in the house, then back home without being caught by the cat.

The aim for the cat is to catch the mouse. If you both land on the same square, the cat has won because it eats the mouse.

▲ Game 2

You could try the game with more than one mouse – or more than one cat!

Three bears

■ Game 1

You need: 1 player
the game board
all the cards, shuffled

Match the cards to the board.

▲ Game 2

You need: 2 players
the game board
all the cards, shuffled
9 red, 9 blue counters

One player has the red counters and the other has the blue counters.

Put the cards in a pile face down. Take turns to turn over a card.

Put one of your counters on the matching board picture.

The winner is the first player to get a line of 3 counters horizontally, vertically or diagonally.

● Game 3

You need: 2 players
all the cards, shuffled

Deal out the cards to the two players.

The first player puts down a card face up in front of them.

Take turns to put down a card with one attribute in common with the last card played, giving a reason for doing so. For example, both cards may show a bed or both may show an object belonging to baby bear.

If you cannot find a suitable card then the other player goes.

The winner is the first player to get rid of all their cards.

■ Game 4 (Extension)

This game is played like game 3, but the card put down must have nothing in common with the last card played.

▲ Game 5 (Extension)

You need: 2 players
the game board for reference
all the cards, shuffled
12 unifix or other small objects

Spread out the cards, face up.

The first player chooses any number of cards. They must have something in common, for example, all belong to Mummy bear, all have 4 legs.

The second player has to suggest the rule for them all being in one set. If they do not give the first player's reason, this second player collects one unifix.

Continue finding rules until you find the first player's rule.

The second player takes a set from the remaining cards and the first player tries to find the rule.

The winner is the player with the least number of cubes when all the cards have been used.

Monster's quilt

■ Game 1

You need: 2 or 4 players
the game board
a dice labelled 3, 3, 4, 4, 5, 6
a different coloured counter for each player

Each player puts a counter on one of the shapes marked S.

Take turns to throw the dice. If you can, you must move your counter to a shape with the number of sides shown on the dice, and with a side in common with your current shape. For example, if the dice shows 3, you may move to a triangle if there is one next to your current shape. If there is no such shape, you can't move.

The winner is the first player to land on an F.

▲ Game 2

You need: 2 players
the game board
10 yellow, 10 green counters

Take turns to describe a shape, for example, 'The shape I'm thinking of has three sides (is a triangle) and is green.'

The other player has to point to a suitable shape.

If your chosen shape matches the description you were given, you can cover the shape with one of your counters.

It is now your turn to describe a shape.

The winner is the first player to get rid of all their counters.

Bugs

■ Game 1

You need: 1 player
the game board
the cards (without the number cards),
shuffled

Put the cards in a pile face down.

Match each card to the same number and same bug on the board.

▲ Game 2

You need: 2 players
the game board
the cards (without the number cards), shuffled
a ten-sided dice (labelled 0, 1, 2, 3, 4, 5, 6, 7, 8, 9)

Deal out the cards equally. Place them all face up so that they can be seen.

Take it in turns to throw the dice. Put a card on the board to make a whole bug whose spots add up to the score. If you can't go, miss that turn.

The winner is the first player to get rid of all their cards.

● Game 3

You need: 2 players
all the cards, shuffled

Pick a card at random and put it face up between the two players.

Deal out the rest of the cards. Place them all face up.

The player with the extra card starts. Take turns to place a card down on the middle pile, matching the previous card in either the number or type of bug.

The winner is the first player to get rid of all their cards.

■ Game 4

This game is played in the same way as game 3, but the rule for putting down cards is changed. For example, the next card must have nothing in common with the last, or the numbers and/or dots must add to 5.

Necklaces

■ Game 1

You need: 2 players (or 4 for a shorter game)
the game board
a dice labelled 0, 0, 1, 1, 2, 2
12 red, 12 blue, 12 yellow, 12 green
counters

Choose two necklaces for each player (one each for four players).

Take it in turns to throw the dice. Cover that number of shells on your own necklace(s).

The winner is the first player to cover all of their necklace(s).

▲ Game 2

This game is played in the same way as game 1 but choose two types of shell instead of two necklaces.

The winner is the first player to cover all their chosen shells.

● Game 3

You need: 2 players
the game board
a dice labelled 1, 1, 2, 2, 3, 3
48 counters

Choose two necklaces for each player.

Take it in turns to throw the dice. Cover that number of identical shells on your own necklaces with counters. If there are not enough shells left on either of your necklaces, you can't go.

The winner is the first player to complete both their own necklaces.

■ Game 4

You need: 2 players
the game board
a dice (1–6)
48 counters

Choose two necklaces for each player.

Take it in turns to throw the dice.

Split the number shown on the dice into two numbers and cover shells from different necklaces with each number. For example, 5 on the dice could be split into three shells on one necklace and two on another.

The winner is the first player to complete their own necklaces.

These games can be played several times in succession. Covering fish in the border could be a way of recording how many games each player has won.

Race to the honey

■ Game 1

You need: 2, 3 or 4 players
the game board
a counter for each player
a dice (1–6)

Start with the counters on the bears.

Take it in turns to throw the dice and move that number of spaces.

If you land on a green space, move forward 10 spaces.

If you land on a blue space, move back 10 spaces.

The winner is the first player to get to the honey. You must throw the exact number to finish.

▲ Game 2 (Extension)

Play this game like game 1, but this time you must calculate in your head where you will end up when you move back or forward 10, and say which space you will land on before you move your counter.

If you are right you can choose whether to move.

If you are wrong you move back 10 spaces.

● Game 3

Make up your own rules for what you do if you land on a red square, or for the green and blue.

Shopping

■ Game 1

You need: 2 players
2 game boards
the cards
a dice (1–6) for items of 30p or over or
a dice labelled 0, 1, 1, 2, 2, 2 for cards
with 10p or 20p items
25 one pence, 10 ten pence coins

Players decide what value of item they will both buy. Each player places a corresponding card on their board at the top of the 10 pence column, so the correct number of 10p coins show.

The aim is to have enough money to 'buy' the item on your card.

Take it in turns to throw the dice. You win that many 1p coins. Put them in the 1p column on the right of the board.

When you have 10 pennies you can exchange them for a ten pence coin.

The winner is the first player to get to the value of their shopping item, so to buy the dinosaur you must win 30p.

▲ Game 2 (Extension)

Play this game just like game 1, but you must tell your partner exactly what you are doing. So if you throw a 6 you might say something like this.

'I have thrown a 6 so I am putting six pennies in the pennies column. I have now got eleven pence there. I am exchanging 10 pennies for a ten pence piece and putting that in the ten pence column.'

Dominoes

■ Game 1

You need: 1 to 5 players

domino cards, either the set with dots and calculator numbers or the set with words and numerals, shuffled

Share out all the dominoes. The player with double six starts.

Domino cards can be laid at either end of the line. Take it in turns to place a domino that matches, so a six is put next to the six already down. It does not have to be in the same form.

The winner is the first player to get rid of all their cards.

▲ Game 2

Play this game as game 1, but use all the cards.

● Game 3 (Extension)

Play this game as game 1, but make the dominoes add up to a number you choose, say eight, so if you start with double six, the number to put next to the six is a two.

What's the time, Mr Wolf?

■ Game 1

You need: 2, 3 or 4 players
the game board
1 set of cards, shuffled (You could use all the
cards or just one type, for example, those
showing digital times.)
a counter for each player

Players start with their counter on their house.

Put the cards in a pile face down.

Take it in turns to turn over a card. Tell everyone
what it says, for example, 'one o'clock'.

If that time is the space next to them, they can
move onto it.

If not, the card is placed on a second pile which is
shuffled and re-used when the first pile is empty.

Players must pick up '1 o'clock', '2 o'clock' etc. in
that order.

The winner is the first player to get to 12 o'clock
and call 'dinner time!'

Snakes

■ Game 1

You need: 2 players
the game board
a dice labelled 5, 6, 7, 8, 9, 10
10 yellow, 10 green counters, one colour
for each player

Take it in turns to throw the dice.

Put one of your counters on the head of a snake having that number of spots. If all the snakes with that number of spots already have a counter, you can't go.

The winner is the player covering the most snake heads.

▲ Game 2

This game is similar to game 1 but you cover a head if the number of spots is 1 *more* than the number on the dice. If you throw a 10 that is lucky – you can have another throw or choose a 5 snake to cover. Alternatively cover a head if the number of spots is 1 *less* than the number on the dice. In this case a throw of 5 is lucky – you can choose to have another throw or cover a 10 snake.

● Game 3

You need: 2 players
the game board
a dice labelled 0, 1, 2, 3, 4, 5
10 yellow, 10 green counters

Take it in turns to throw the dice.

Put one of your counters on a snake head if the number on the dice and the number of spots on the snake *add up* to 10.

The winner is the player covering the most snake heads.

■ Game 4

You need: 2 players
the game board
a dice labelled 0, 1, 2, 3, 4, 5
10 yellow, 10 green counters
20 unifix or other small objects

Take it in turns to throw the dice and collect that number of cubes.

If you have enough cubes to cover all the spots on a snake, you can put one of your counters on the head of that snake and put that number of cubes back in the middle. Keep any cubes left over to use on your next turn.

The winner is the player covering the most snake heads.

▲ Game 5

You need: 1 or 2 players
a dice labelled 0, 1, 2, 3, 4, 5
a copy of page 19 for each player
crayons

Take it in turns to throw the dice.

Colour in that number of spots on one snake only.

You must throw the exact number to finish a snake.

The winner is the first player to colour all the spots on their sheet.

Snakes

Introduction to the mats

Philosophy and aims

The mats are designed to provide a stimulating resource for young children which may be used in a variety of ways to develop their mathematical understanding and to provide opportunities for the development of practical work and discussion. They can be used very flexibly to support various teaching styles and abilities of children.

The mats provide practice and reinforcement on all areas of the mathematics curriculum. Each mat has been designed with a particular mathematical concept in mind, but many have more than one concept featured. This allows you, the teacher, to select what appeals to you and your children. Each mat has a clearly defined border depicting a particular feature or concept, many of which can be used for playing games or carrying out simple investigations.

The mats can also be used for assessment and review. The Teacher's resource book gives detailed advice on how six of the mats can be used in this way, but many of the others can be used equally as well.

The main emphasis in this pack is on activities which are suitable for use with four or five year old children, but suggestions are also included for further development work and extension activities to show how they can be used with a wider age range if so desired.

One copy of each mat is available in this pack. It is often useful to have further copies so several children can work on the same one at once. The Extra mats pack enables you to buy as many extra copies as you wish.

How to use this book

Within this book there is a section on each mat. These outline some of the key concepts which may be introduced using the mats and list the suggested resources which may be used for practical work. Restrictions on space mean that it is only possible to give a summary of possible activities. The sections below give some ideas on how these can be expanded to suit your children and ways of working. Some mats have sample resource sheets which can be photocopied for recording and follow-up tasks.

How to use the mats

The same principles may be applied to any of the mats and these are outlined below. It is important always to start with discussion and this can lead to practical activities, investigations and problem solving, simple games, assessment and review, and recording.

Discussion

Discussion is the most important aspect of the work. It is very important throughout to give the children plenty of time to think and respond to questions. Ask questions of each member of the group to ensure that all children take part in the discussion.

Start by asking open-ended questions such as,

> 'What can you see on this mat?'
> 'What shapes can you find?'
> 'What do you like about this mat?'

After this initial observation of the mat, the questions can focus on more specific areas. For example, with the Teatime mat, you could ask questions such as,

> 'Which of these meals would you like for tea?'
> 'What is your favourite food?'
> 'What did you have for tea yesterday?'

Follow up responses with further questions to focus upon detail, for example,

'Why do you think that necklace is the longest?'
'How can you check?'

Practical activities

Each mat is designed to stimulate a range of practical activities. Many ideas are outlined in the notes for each mat, but here are some general activities which can be applied to most mats.

- ■ 'Point to the ... '

- ▲ Matching and sorting using appropriate resources.

- ● Counting using counters or objects. Making comparisons.

- ■ Measuring using string, ribbon, etc. Making comparisons.

As part of the reinforcement process, it is a good idea to follow the practical activities with some form of recording.

Recording

Children need to be introduced to the idea of recording, but it may actually be done by the teacher or an adult helper after discussion with the child. The advantages of this are that children

- ■ begin to experience different ways of recording and can compare them,

- ▲ are encouraged to consider what the key points are that need to be recorded,

- ● begin to see the value of recording as a reference for further work and as a means of communication,

- ■ begin to see a logical approach and organisation to recording,

- ▲ are encouraged to record when they are ready and not before, so children's growing confidence is not harmed.

Children can record their ideas using a wide range of media. Often the only recording valued is pencil and paper and this is unfortunate as it gives a very narrow impression of how we learn and communicate mathematics. It is possible to record on the mats themselves by using a washable ohp pen. The recording could also be drawing, colouring, matching, filling gaps on mats resource sheets, painting, printing and model making. These may require some adult supervision. Sewing and cooking are also very worthwhile, but require a higher level of adult supervision from parent helpers or ancillary staff.

Stimulating investigation and problem solving

The mats are carefully designed to provide opportunities for open-ended investigation and a stimulus for problem solving. Some ideas are included in the notes for each mat, but there are some general features to look for when preparing work that may be applied more generally.

The introduction of free and structured investigation of mathematics at an early age helps children to develop a broad insight into the fact that there are often many ways of solving a problem. It is important to encourage children to develop confidence in exploring a variety of approaches to problem solving and to develop strategies for investigation.

Simple games

Each mat has a border which provides opportunities for developing games. Some ideas are provided in the notes, but you can also make up your own. These may be played by children in groups of up to four or as a teacher-directed whole-group activity. Children should also be encouraged to develop their own games using the mats and equipment such as counters and dice.

It is important to discuss with children the basic principles of playing games. These are the rules (especially start and finish rules, missing turns, extra turns, time limits set, frequency and order of play), aims (how the game will be won), and scoring (this should be kept simple, a calculator may be useful).

Assessment and review

A particularly valuable feature of the mats is that they may be used to monitor progress using practical activities, especially if several copies of each mat are available. The Teacher's resource book includes review activities based on the Dice and dominoes, Clowns, Teddy collection, Snakes, Sweet shop and Jewellery mats. However it is anticipated that any teacher using the mats will be making continuous assessments of the children as they carry out activities and further work will be based on these assessments. It is important to give children time to consolidate their understanding of concepts before moving on to higher-level skills.

Classroom practice

It is recommended that mats are used daily in ten-minute sessions. However even five minutes' discussion can provide a wealth of mathematical ideas and vocabulary, and some children may be able to concentrate for 15 or 20 minutes while exploring an investigation together. It is expected that each mat will be used several times with each group and that on each occasion a particular aspect of the mat is used as the focus. An example of how this can be done is given in the next section.

Mats can be used with the whole class, groups, pairs and individuals. They should also be freely available for children during free choice or informal activity sessions as they can stimulate imaginative play and discussion.

Whole class use

Mats can be attached to an easel and used for general class discussion or displayed as a poster on the wall for reference. They can be useful as a general introduction to an appropriate theme or topic, for example the Teatime mat is good for introducing the topic of food, the Sweet shop mat for money and the Teddy collection for toys.

Group work

Groups of up to eight children could work on a single mat around a table with the teacher or ancillary. Ideally have several copies of the same mat available so children can work in pairs or individually on the practical activities. This enables all children to be actively involved and makes assessment easier.

An example of daily ten-minute sessions using the Dice and dominoes mat

Session 1	General observation and discussion of all aspects of the mat	**Session 8**	Dominoes and number line: matching
Session 2	Dice only: matching and sorting using actual dice		
Session 3	Border, number line to 10: match, group counters, cover and hide numbers, number rhymes, more than and less than	**Session 9**	Dice: combinations and totals
Session 4	Dominoes: sorting, matching and counting		
Session 5	Dominoes: grouping into sets	**Session 10**	Dominoes: making 10
Session 6	Dominoes: matching pairs		
Session 7	Number line to 20, extension of session 3		

Paired activity

Children can be set to work together on a mat as an introduction to an activity or game or as a reinforcement for other work.

Individual work

Mats can be used to teach a new concept to an individual child or to provide an activity for a more or less able pupil. They can also be used for reinforcement for an individual child.

Grouping children according to ability

It is possible to adapt activities so that they are suitable for different abilities. This example shows how the Sweet shop mat can be used with three groups of 6–8 children grouped according to mathematical ability. It is unlikely to be this simple at reception level, but the example shows how a range of questions can cater for a wide ability range. Similar activities could be done on this mat based on measures and money.

Lower ability group	Average group	More able group
Use jars 2 and 4. Discuss the contents. Compare/count/match to decide which jar has more. Introduce jar 3. Discuss the contents. Provide each child with a jar shape cut out of paper. Use counters or sweets for each child to make a collection of 9 sweets. Use 2 or 3 colours to enable the introduction of addition, comparison (more/less) and sorting. Spend time talking about the contents of each child's jar.	Use jars 1, 3 and 5. Discuss the contents. Count/compare/match. Introduce more than, less than, most and least. Look at the details of coloured sweets. Order the three jars according to the number of contents.	All jars. Ask open-ended questions such as, 'Tell me everything you can about the jars.' Go around the group and record each comment on a flip chart or list. Extend vocabulary to include 'greatest number of'. Begin to consider totals of various shapes of sweets, sizes of sweets. Order the jars according to the number of contents.

Dice and dominoes

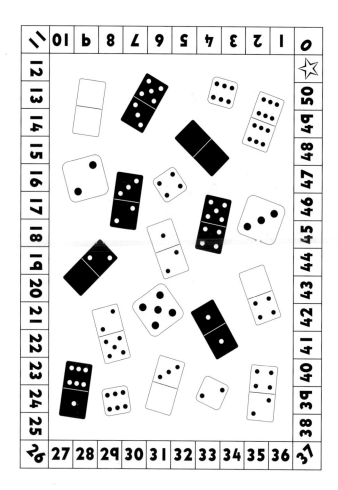

The border provides a multipurpose number line displaying the numbers 0 to 50. A star marks the start and finish of the line which may be used for a variety of games. These spaces are large enough for counters or multilink. The mat can be used for any activity requiring a simple number line.

Useful resources

Selection of dice, variety of dominoes, counters, dominoes resource sheet, dice resource sheet

Teacher's notes

Using the dice

■ **Matching with real dice**. Begin by matching actual dice onto the dice on the mat so that the correct numbers of spots are shown. Ask lots of questions such as:

'How many spots are showing on the large blue dice?'

'Which dice has the least number of spots shown?'

'Are any of the dice the same? Why?'

▲ **Adding the spots on two or more dice**. Ask children to total the number of spots on various dice shown on the mat.

'What is the total number of spots on the blue dice?'

'What is the total number of spots on the three large dice?'

'Find two dice with a total of 8 spots.'

'Find two dice with an odd total number of spots.'

● **Adding spots on real dice.** Explore totals scored when rolling pairs of dice.

Gradually introduce more dice – three, then four.

Mathematical concepts

Number recognition, sorting and ordering, matching, number bonds, logic

Description of the mat

The design uses various examples of dominoes readily available. These include:

the standard white spots on black background set, the black spots on white background set, and the coloured spots on black background set.

This mix of dominoes has been used to provide variety and lead to a range of different investigations. Blank dominoes have also been included.

Suggest that children record their results as they roll the dice.

For example:

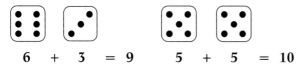

6 + 3 = 9   5 + 5 = 10

Children can use counters or bricks to help them.

6   +   3   =   9

■ **Comparing spots on dice.** Compare scores when rolling a pair of dice. Which has more spots? How many more are there?

Using the dominoes

■ **Matching with real dominoes.** Using a set of dominoes, find dominoes to match each domino on the mat.

▲ **Adding the spots on a domino.** Discuss the values of the various dominoes when you total all the spots:

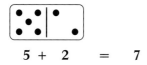

5 + 2   =   7

● **Sorting real dominoes.** Using a standard double-six set of dominoes, sort the dominoes in as many ways as possible:

'Can you find a domino with 8 spots?'

'Find all the dominoes with 5 spots.'

'Find all the dominoes with more than 6 spots.'

Take a handful of dominoes and order them according to the number of spots on each. 'Which domino has the most spots?'

■ **Building towers to match domino spots.** Build towers of bricks or cubes to match the number of spots on the dominoes. Put together the two cube towers on each of the dominoes. Do any of the dominoes have the same total number of cubes?

Which dominoes together would have a total of 11 cubes?

These could be recorded:

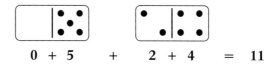

0 + 5   +   2 + 4   =   11

Generate other similar simple investigations using the dominoes in this way.

Those who are more advanced could be asked to record sets of 3 dominoes that have a total less than, say, 20.

Add a tower of cubes to each domino to make a total of, say, 12.

Add a tower of 4 cubes. $3 + 5 + 4 = 12$

▲ **A guessing game.** Tell the children that you have chosen a domino on the mat and they must guess which it is. Give clues or invite the children to ask questions (requiring the answer 'yes' or 'no') to find out exactly which one you have chosen.

'This domino has white spots on both halves.'

'This domino has 7 spots.'

Using the dice and dominoes together

Once the children have developed confidence in using both dice and dominoes for a variety of activities, the two may be used together.

■ **Matching real dice and dominoes.** Use a box of dominoes and a single dice to begin with. Roll the dice, look at the score and record it.

Find all the dominoes in a box with this number of spots.

For example:

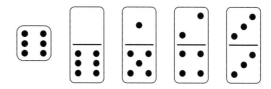

Roll 2 dice and again find dominoes with the same total number of spots.

For example

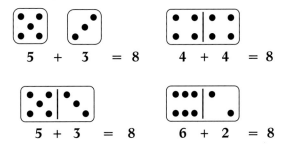

5 + 3 = 8   4 + 4 = 8

5 + 3 = 8   6 + 2 = 8

▲ **A game matching dice and dominoes for up to 4 children.** The children take it in turns to roll a dice (1–6). Each child then chooses a domino from the box with this number on at least one side.

For example, roll

Collect any domino with 4 spots on one side, for example,

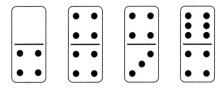

Play continues until all the dominoes have been collected.

The player with the most dominoes at the end is the winner.

Alternatively, introduce a calculator and invite the children to total the number of spots they have collected. The winner has the highest score.

● **Children's own games.** Ask the children to devise their own games using dice and dominoes.

Using the border

■ **Matching dominoes to the number line.** Using a complete set of dominoes, find the total number of spots on each and place it against that number on the border. Which number has the most dominoes?

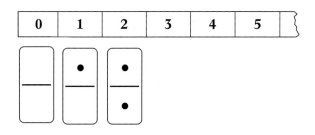

▲ **A game for 2 or 3 players.** Place a set of dominoes face down in the centre of the mat. Each player needs a different coloured counter placed on 0 to start the game.

Players take it in turn to pick up a domino and count the number of spots. They move their counter round the border that number of spaces.

For example:

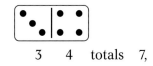

3   4   totals   7,

so the player moves 7 spaces.

Once used, dominoes are put in the box. Play continues until all dominoes are in the box, or until a player reaches 50.

Adapt the game according to the counting skill of the children. For example, you could remove all the 5 and 6 pieces, creating a double-four set, involving numbers up to 8 only.

This game will begin to introduce the numbers up to 50.

● **Introducing odd and even.** Use the coloured backgrounds in the border to introduce odd and even. Give children odd and even numbers of counters and encourage them to discover that it is possible to share even numbers between two people.

Provide children with cubes or counters and ask them to cover over the odd numbers less than 10, say. Read out all the even numbers. Extend to 20 and ask the children to write down the even numbers.

The dice resource sheet

Once you have worked with the children on a practical activity, the dice resource sheet can be used for recording and extension. This sheet is open ended so that you can adapt it for an activity appropriate to the group or individual. Mark up a copy before duplicating it for the group. A collection of assorted dice would be useful. The notes describe activities using a standard 1–6 dice, but they could all be extended to use 1–10 dice or others.

As the children become more confident with dice activities they may suggest recording ideas of their own using these sheets or suggest other sheets. Do encourage them.

- **Recording and comparing numbers on dice.** Use a 1–6 dice and part 1. Roll the chosen dice and write the score into one of the squares. Continue until all squares contain a number.

  Use colours to sort the numbers. Colour 1s orange, 2s blue, 3s red, etc.

  Alternatively divide the sheet into 3 rows. In each section colour the highest number red and the lowest yellow. (If there is more than one, colour all of them.)

  For example, in row 1, the 6 would be coloured red, the two 1s yellow.

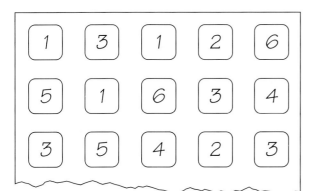

- ▲ **A game recording totals on two dice.** Use the sheet with two 1–6 dice to challenge the children to plan their recording. Divide the sheet into 5 columns of three squares.

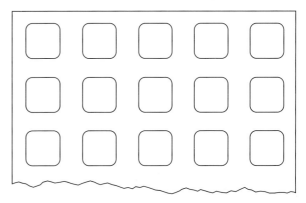

Throw both dice, and add the scores. Record the total in any square, but at the end of the challenge each column should contain 3 different numbers. This may be played individually, with a partner or as a group.

- ● **Recording totals on three dice.** Use part 2 to record three throws of a dice and their total. Calculations can be recorded beneath each triple as in the example below.

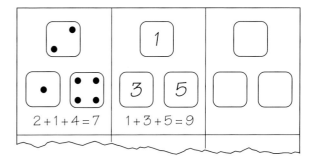

Record with spots or numbers.

The domino resource sheet

This sheet can be used in similar ways. A collection of assorted dominoes may be useful.

- **Drawing a given number of spots.** Write a number from 0 to 12 under each domino. Children then find a domino with the appropriate total and draw in the spots.

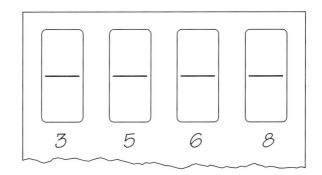

▲ **Cut, sort and stick**. Use the sheet to cut and stick. For example, draw spots on the dominoes so that they can be put into pairs with the same total. The children cut them out, sort them into pairs, and stick them in their pairs on paper or on a sheet divided into 6.

Paper dominoes can also be used to record other work children have done with real dominoes.

Support activities

■ **Matching spots with towers of cubes.** Ask children to build towers of bricks or cubes to match the number of spots on specified dice.

▲ **Matching real dominoes.** Children can match the ends of real dominoes like this.

They could go on to make a pattern such as:

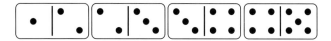

● **Counting spots of various colours**. Using the mat and a box of dominoes with coloured spots talk about the various numbers of spots for each colour.

Extension activities

■ **Introducing numbers to 50.** Talk about the numbers in the border, looking for numbers that the children can recognise:

(30) 'This is the number of my house.'

(5) 'This is how old I am.'

(42) 'This is the number on the bus when I go to gran's house.'

Make a set of cards with number words (two, twenty, thirty-five, etc) and ask children to match a selection of these words with the numerals in the border, trying to give each number a name. Gradually introduce the full range of numbers, but start with a limited selection.

▲ **A missing numbers game.** When the children are not looking, use dominoes or large counters to cover a few of the numbers in the border. Ask the children to say what the hidden numbers are.

● **Finding numbers to satisfy a rule.** Ask the children to cover a selection of numbers using large counters or cubes, for example:

all the odd numbers;

all the even numbers less than 20;

all the numbers between 10 and 20;

any number that contains a 4 digit (4, 14, 24, 34, 40–44, 49);

any number ending with a zero (10, 20, 30, 40, 50);

numbers whose digits total 4 (4, 13, 22, 31, 40).

Ask children to suggest their own rules.

■ **Finding missing spots on the dominoes sheet.** Write a total (0 to 12) under each domino, and draw some of the spots. Children count the spots and insert the missing spots to make the total. Recording could be extended by putting the total above and the sum below.

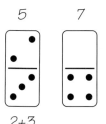

▲ **Finding dominoes with a given number of spots between them**. Draw lines on the dominoes sheet to split the dominoes into 6 pairs. Write a total (1 to 23) above each. The children find pairs of dominoes that give the required total, and draw in the spots.

Alternatively the grid may be divided into 3s by vertical lines or 4s with horizontal lines and children could find dominoes with the totals given. Encourage the use of a calculator:

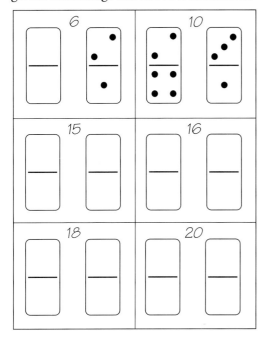

Find three dominoes each having 6/9/10/11 spots.

Find 4 dominoes with a total of 15 spots.

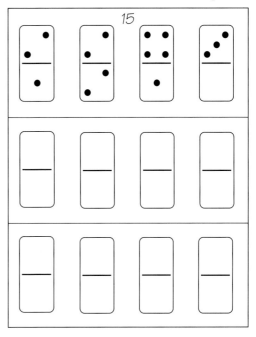

● **The dominoes sheet for practice work.** Use a sheet divided into six pairs. Write numbers in one domino of each pair before copying. The children put in the correct number of spots on the second domino.

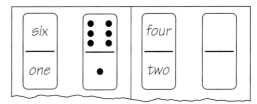

Alternatively simple addition or subtraction could be written in the first domino.

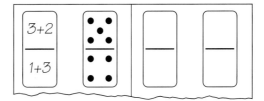

Additional information could be written in between (such as the answer) if the children need practice in writing numerals.

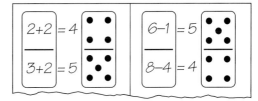

You could reverse this process by drawing in the spots and asking the children to write an appropriate calculation in the second domino.

Dice

Dominoes

Name...

Triangles, circles and squares

Mathematical concepts

Shapes, matching and sorting, number, number bonds, pattern, introducing multiplication

Description of the mat

There are 6 adjoining 12 x 12 cm squares. Within each is an equilateral triangle of side 11 cm, containing 3 circles. The circles can be used to introduce children to multiples of 3 in a practical way.

The colours have been selected to enable counting and comparison to take place as part of the general discussion.

The border is divided into 4 sections. The top and bottom each contain 12 white circles, the sides each contain 10 circles – 44 circles, in all. These can be covered with standard counters, and thus used for a variety of counting activities or games.

Useful resources

Circles, squares and triangles, counters, cubes, dice; geostrips, straws and pipe cleaners, Polydron or Clixi, sticky paper shapes, shape stencils, kite resource sheet, and copy of the mat, which may be used for recording purposes

Teacher's notes

- **General discussion.** Begin with a general discussion about the mat. Count triangles, squares, circles. Discuss the use of colour.

- ▲ **Introducing subtraction.** Match coloured counters onto the circles and count them. Play games, removing counters and counting the number left behind. This will introduce the idea of subtraction.

- ● **Identifying a section from a description.** Describe a part of the mat and ask the children to point to the section you have described.

  'This section has blue circles on a yellow triangle.'

- ■ **Describing a hidden section.** Cut out pieces of card 12 x 12 cm. Use these to cover one of the six sections of the mat, then challenge children to describe exactly what the design is under the card.

  'I think it is a blue square with a red triangle with a yellow circle in each of its corners.'

- ▲ **An investigation.** Investigate ways of colouring the sections using the same colours that would be different from any of the six shown.

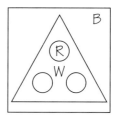

- ● **Introducing multiplication.** Ask children to investigate the totals if the same number is put into each circle in any one triangle, like this.

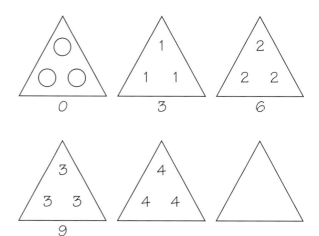

Triangles

Name..

NCM Module 1 Mats

Kites

Name...

Using the border

- ■ **A race game for 2 to 4 players**. You need a standard 1–6 dice and a coloured counter for each player. Choose a starting point at the beginning of one of the sides. Take it in turns to roll the dice and move around the appropriate number of places. The winner is the first player to get back to their own starting place. The exact number is needed to land on the circle for the final turn.

- ▲ **A game for 2 to 4 players collecting counters**. You need a standard 1–6 dice, about 15 counters all of the same colour, a different coloured counter for each player.

  Put the 15 counters around the border on the circles at random. All players put their counters on the same starting place. Take it in turns to roll the dice and move around the border. If you land on a counter then you may collect it. The winner is the player who collects the most counters.

The kite resource sheet

- ■ **Splitting up a number**. Give the children 10 counters or cubes and ask them to split these up in various ways, for example into 6 and 4. Ask the children to write their pair of numbers in a kite. Encourage them to find other ways to do it.

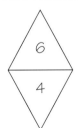 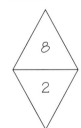 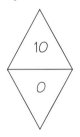

  Children could split up other numbers according to their ability.

- ▲ **An investigation into the number of ways to colour a kite**. Give children triangles of four (or fewer) different colours: red, yellow, blue, green. Show them how to make a kite by putting two triangles together.

  Ask them how many different kites they can make. Encourage them to use the resource sheet to record their designs.

Related activities using other resources

- ■ **Patterns from triangles**. Cut out a set of triangles using card, paper or felt. Invite children to make patterns using the pieces.

- ▲ **Making triangles**. Make triangles using geostrips, card, or straws and pipe cleaners.

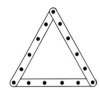

- ● **Building shapes from triangles**. Build 2D and 3D shapes using triangles from construction materials such as Polydron or Clixi.

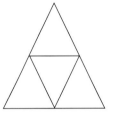 **2D**

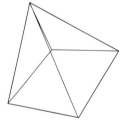 **3D**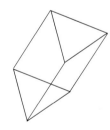

- ■ **Making designs with triangles and circles**. Ask children to make up some designs using 2 triangles and 6 circles. These could be made with sticky shapes, or drawn using the stencil.

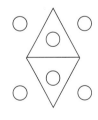 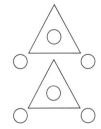 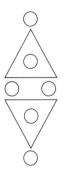

Extension activities

■ **Making a target number with real objects**.
Choose a target number, for example 12. Build
towers of bricks or counters on the circles
within each triangle so that the total in each
triangle is the same but no two triangles have
the same towers. For example:

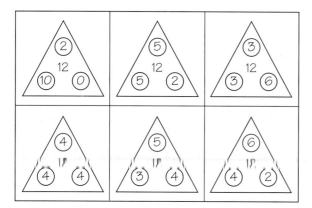

This is a good group activity.

▲ **Adding three numbers**. Use a copy of the mat.
Write numbers in the circles for the children to
total and write the answer in each triangle, or
invite the children to write their own numbers
into the circles, perhaps restricting their choice.

$$7 + 3 + 1 = 11$$

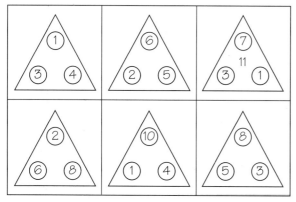

Alternatively, put a number in the middle and
invite the children to find ways of placing three
numbers in the circles that will add up to the
middle number.

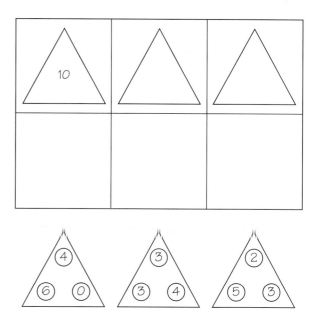

could all complete the first triangle.

Clowns

Mathematical concepts

Shape, number matching, pattern, comparison, size, investigation, data handling, sets (partitioning), addition and subtraction

Description of the mat

This mat has been designed to promote mathematical discussion, close observation and comparison. Within the central section there are four clowns, each different in height. There are many details to be identified and compared.

Clown 1 is the shortest, with middle sized feet, laced shoes and no hair. He has 4 pockets, no patches and 1 hexagonal button on his jacket. He is wearing a bow tie, a flat hat with 3 feathers and is happy.

Clown 2 is slightly taller than clown 1. She has long orange hair, and a top hat. She has no pockets, 4 patches, and 2 round buttons. She is happy and looking up as if ready to juggle.

Clown 3 is slightly taller than clown 2, and is looking down. He has 1 pocket containing a pen and pencil, 2 round buttons, 3 patches and the smallest feet. He is the only sad clown, with bare arms, wearing glasses and a pointed hat.

Clown 4 is the tallest. He has 2 pockets on his jacket, 2 pockets on his trousers, 2 patches on his trousers and 2 elbow patches on his jacket. He has 4 round buttons on his shirt and 4 triangular buttons on his trousers. His feet are the biggest and his shoes have laces. He also wears a bow tie, and a top hat with a flower. He is happy and has hands held high as if to juggle.

Patches use only 4 types of fabric to enable matching and comparison:

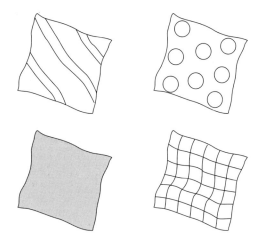

In the border there are 9 skittles (3 each of 3 different kinds), 8 sticks (2 each of 4 different kinds), 6 balls (2 each of 3 different kinds) and 10 flowers (1 green, 2 yellow, 3 red/pink and 4 blue).

Useful resources

Counters, skittles, sticks, balls (if possible), clowns resource sheets

Teacher's notes

The clowns

Start with lots of discussion about the clowns. This should include counting items such as pockets and comparing sizes.

■ **Counting cubes given to each clown.** Using identical-sized counters or cubes the children can give clowns various items to juggle with. Give them instructions and refer to the border:

'Use counters to cover and count the skittles in the top border. Clown 4 is juggling with these skittles. How many does he have?'

'Use counters to cover and count the balls in the border (6). Clown 2 is juggling with the balls. How many does she have?'

You can also select using colours as the criterion. For example,

'Clown 3 has 3 red flowers, 4 blue flowers, 2 yellow flowers and 1 green flower. How many are there altogether?' This example has used all the flowers available, but obviously it is not necessary to do so.

▲ **Matching shapes.** The corners contain shapes. These can be used to help children find the buttons which are triangles, for example.

Using the border

■ **Counting and matching.** There are plenty of opportunities for discussion, for example,

'How many yellow flowers are there?' (Match with coloured cubes and build towers.)

'Match the pairs of balls.'

'Match the skittles.'

'How many balls are there in total?'

▲ **Quick calculations using real objects.** A set of plastic skittles or coloured sticks can be used to stimulate quick mental calculation with a group or whole class by playing simple carpet games. For example, put out 8 skittles of various colours. 'Sort the skittles and find the number of each colour.'

● **Making patterns using real objects.** Make patterns where children tell you what will come next.

Remove skittles while the children are not looking and challenge them to identify what has been removed.

■ **Counting objects in a clown's bag.** Make a collection of items a clown may have in his bag. This should be done using actual objects if possible.

for example,

3 skittles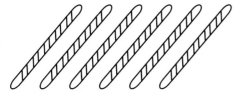

6 sticks

4 balls

13 things

'He takes out the sticks. What is left in the bag?'

Suggest ways of colouring, sorting or ordering the items.

The object cards

The object cards contain 1 to 8 of each of the four items, skittles, sticks, balls and flowers. They are provided on resource sheets and may be copied onto card, cut up and used to develop a variety of activities with individuals or small groups of children.

■ **Sorting.** Ask children to sort the cards into four piles of skittles, sticks, flowers and balls or into eight piles according to the number of objects.

▲ **Playing snap**. Use the cards to play snap where the cards are shuffled and shared between two players. A snap can be for the same item or the same number.

● **Introducing simple multiplication**. Use simple number patterns and counting to introduce simple multiplication, for example,

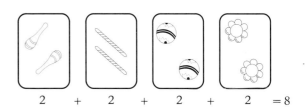

$$2 + 2 + 2 + 2 = 8$$

Follow-up tasks using recording

■ **Producing a bar chart.** Produce a chart showing the items in the border. Draw items or colour squares of the copy of the square grid mat.

▲ **Make your own clown**. Draw or collage a clown with pockets, patches, buttons etc. Talk or write about your clown.

● **Adding using the object cards.** Take 2 or 3 of the object cards at random or select them to produce a particular total. Match with the corresponding number cards, then record.

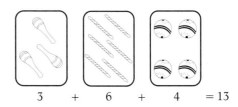

$$3 + 6 + 4 = 13$$

Support activities

■ **Counting objects.** There are ample opportunities for counting objects, for example buttons, pockets and patches on the clowns. Also the children can count how many flowers, say, there are in one part of the border and compare this with the number in a different part.

Extension activities

- **An investigation.** Use only two colours, for example red and yellow. What different arrangements of parcels (cubes) could the clowns carry? For example, with 3 parcels there are 8 possibilities:

R	R	R	Y	Y	Y	R	Y
R	R	Y	R	Y	R	Y	Y
R	Y	R	R	R	Y	Y	Y

These could be recorded on squared paper or on a pre-prepared sheet.

- ▲ **Making ten.** Ask the children to sort the object cards so that they make collections of 10 items, for example,

3 skittles   3 sticks   4 balls

Children can be encouraged to help each other.

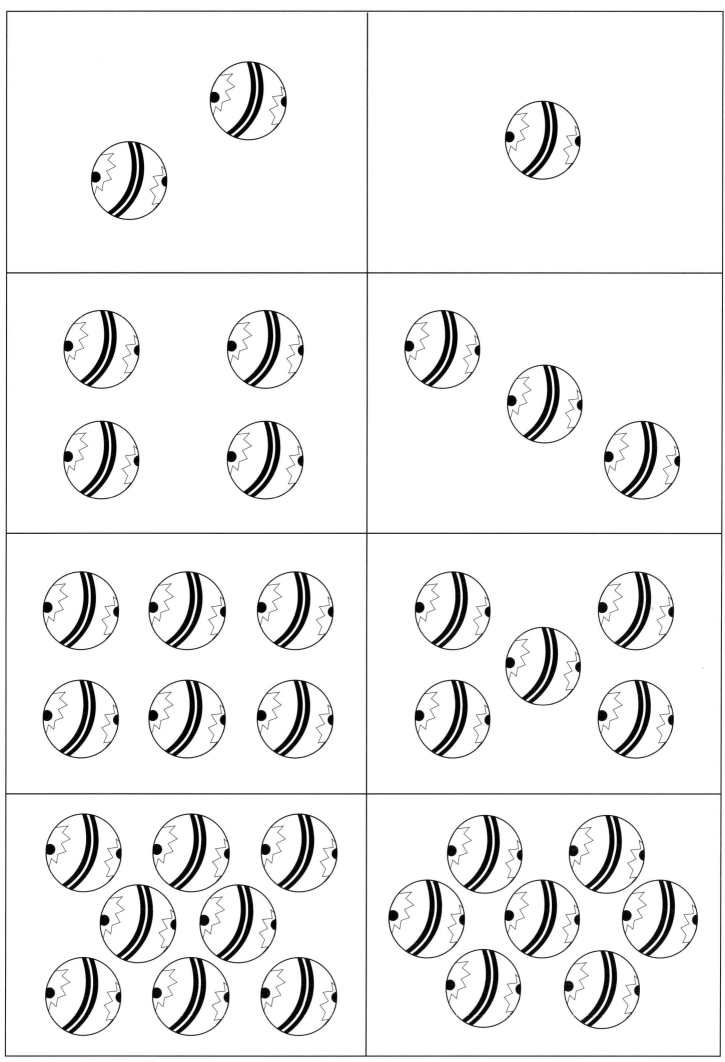

Object cards

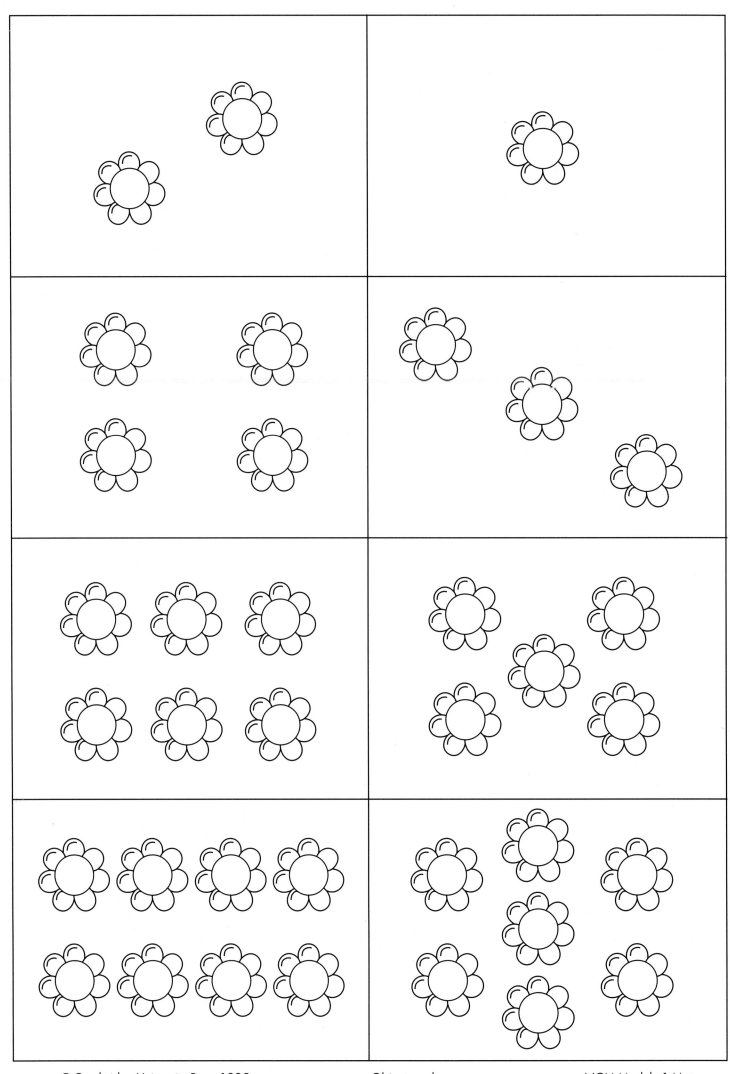

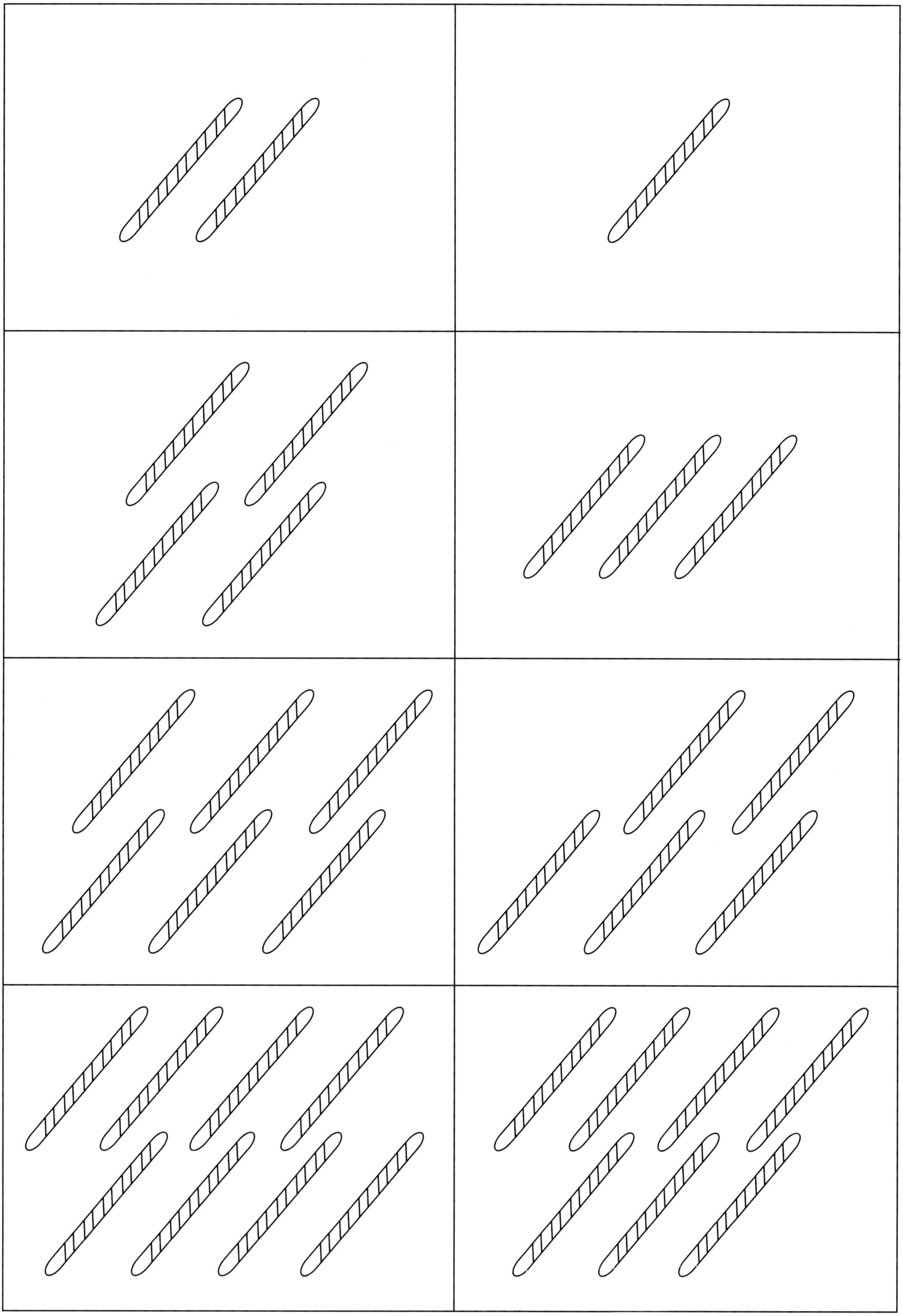

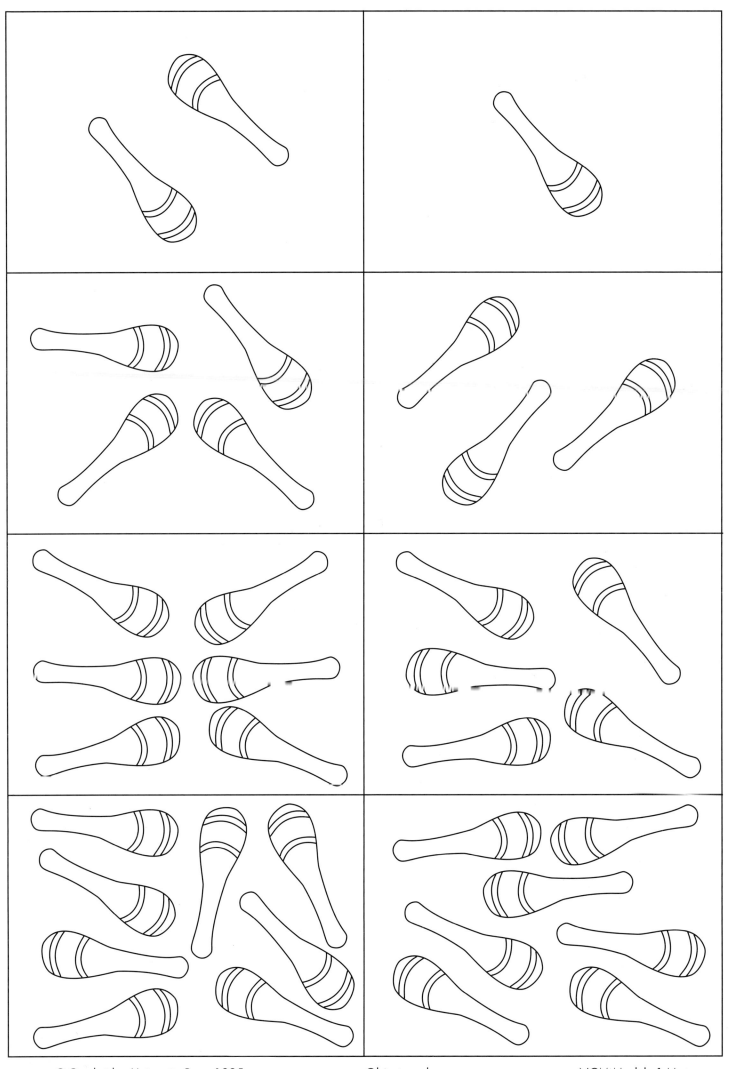

Object cards

Square grid

Mathematical concepts

Number, symmetry, pattern, co-ordinates, data, matching, simple area, logo

Description of the mat

This multi-purpose mat has been designed to allow for maximum flexibility. It can be used in conjunction with the other mats for practical activities and recording. The squares are large enough for counters or cubes to be put into each or for other objects or stickers to be used. It may be used either vertically or horizontally, giving a 10 × 6 or a 6 × 10 grid. There is a black and white version as a resource sheet for recording or for the preparation of activities to be carried out on the grid.

The scalloped border can be used to insert numbers/colours/words using a washable pen or chinograph pencil. There is room to write in labels round the grid to use it as a graph.

Useful resources

Cubes, 1–6 dice, unifix number indicators are useful but not essential, copy of the mat

Teacher's notes

■ **Finding a given number of cubes.** Insert numbers in the scallops for children to collect the appropriate number of cubes. Unifix number indicators are a good way to show the numbers if you have them. If you use the mat vertically you can insert 6 numbers up to 10. If you use the mat horizontally, then you can have 10 numbers up to 6. Children can record on the mat or the resource sheet.

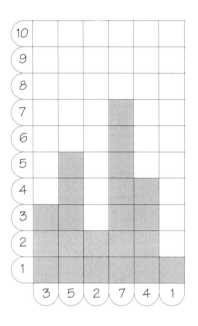

▲ **Adding using cubes.** For addition up to 6 or up to 10, use the grid with two colours of cube to show 2 numbers added. Write the numbers in the scallops at either end.

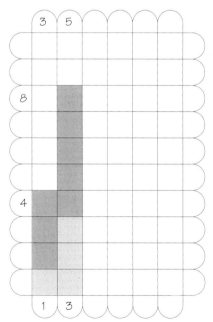

● **Drawing patterns.** The grid is ideal for work on pattern. It can be used vertically or horizontally for longer patterns. Start a pattern for the child to continue.

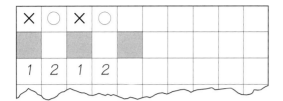

■ **Recording simple calculations.** The rows of the grid provide a recording sheet for simple calculations using calculators. Using colour to record the +, – and = signs helps the child to locate them appropriately.

3	+	4	=	7
2	+	4	=	6

1	0	+	1	2	–	2	2

Playing games

The grid is a useful resource for playing many different games. Here are some examples.

■ **A simple game for two players.** You need a 1–6 dice and plenty of counters or cubes. Write the numbers 1 to 6 in the scallops at both ends. Players take it in turns to roll the dice and record their score by colouring a square or placing a cube or counter at their end of the appropriate row. The first player to reach the middle of the mat is the winner.

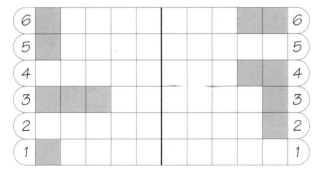

▲ **A racing game for two players.** You need six different coloured cubes or counters, a standard 1–6 dice and a colour dice for the six colours being used. Use the grid upright and put the coloured counters at the bottom of the grid on the six scallops.

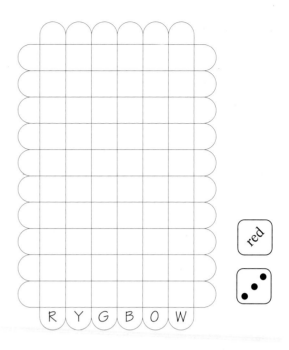

Take it in turns to roll both dice and move the appropriate coloured counter up the grid. For example, with the throw shown move the red counter three spaces. Play continues until a counter reaches the scallops at the top of the grid. The player who moved it to the scallops keeps that counter and play continues. Anyone then rolling that colour misses a turn. The aim is to be the first player to collect three counters.

● **A treasure hunt game for two players.** You need a standard 1–6 dice, 6 red counters, 6 yellow counters and a collection of small objects such as beads, toys, sweets, cubes in a box (the treasure chest).

Mark the grid at random with some numbers appropriate to the children's abilities. Here 1, 2, 3 have been placed in each row.

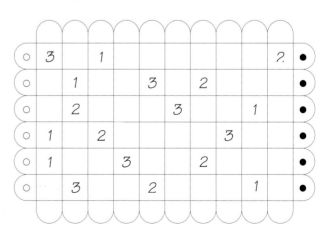

Player A starts at one end with six red counters and player B starts at the other end with six yellow counters. The aim is to move your counters to the other end of the mat while collecting as much treasure as possible.

Take it in turns to roll the dice and to choose one of your counters to move that number of places along the grid. If you land on a number you collect that many items from the treasure chest. For example, with a score of 2, the player starting at the right can move either their first or fourth counter forward two places and collect one item of treasure.

When the counters are near the other end any number exceeding the total number of grid squares left will enable you to move your counter onto a scallop. Once one player has put all their counters on the scallops at the opposite end, the game stops. Players count the number of items of treasure collected.

It is useful to discuss strategy as the children play.

Support activities

■ **Number writing practice.** Use the grid as a recording device for number writing practice. Write in the scallop at one end the number to be written, and at the other end the number of times (up to 6 or 10) the child is to write it. The advantage of this open grid is that you enter only the numbers the child needs to practise.

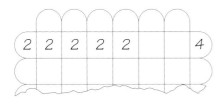

Extension activities

■ **Introducing symmetry**. To provide an introduction to symmetry a design can be produced on one half for the children to reflect in the other using coloured stickers, cubes or crayons. A range of designs can be made to provide for different ability groups.

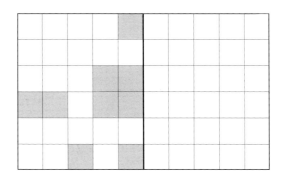

▲ **Introducing simple co-ordinates.** Simple co-ordinates could be introduced as a group activity using the grid marked 1 to 10 vertically, and A to F horizontally. Give instructions to place shapes, colours, numbers, etc.

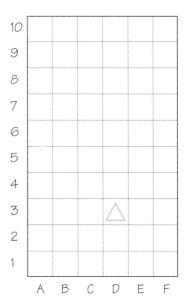

'Draw a triangle in D3.'

'Draw a flower in B5.'

'Draw a circle in F1.'

'Write a number 3 in A10.'

'Colour C2 blue.'

This type of activity enables you to check children's knowledge of shape, colour, number recognition, etc. and is an excellent diagnostic tool.

Square grid

NCM Module 1 Mats

© Cambridge University Press 1995

Teddy collection

Mathematical concepts

Counting, mathematical vocabulary of position, left and right, sorting, data handling

Description of the mat

There are four rows of teddies, each different in some ways. There are also similarities for the children to look for. The teddies are displayed as if on a shelf unit and therefore a wide range of vocabulary may be introduced when discussing their positions.

On the top shelf there are 5 large teddies, 3 standing and 2 sitting. The clown is juggling with 3 balls; the lady with the flowery hat has a small teddy in her arms. One of the teddies is sad.

On shelf two there are 5 teddies, a boxer, a trumpet player, an artist, a flower seller and a teddy who looks very cold. One teddy has closed eyes.

On shelf three, only one of the 6 teddies has a hat on; one has sun glasses. There are two ice creams.

On the bottom shelf there are five teddies. The one throwing a snowball is wearing a hat like the teddy sitting on the top shelf. Four teddies are holding things.

All the way round the border there are items which belong to the various teddies. The children can match and point or match and cover using pairs of counters. The items could be sorted into groups using counters to cover each collection.

Useful resources

Counters, a variety of teddies

Teacher's notes

■ **General discussion.** Some general questions may help children to gather information from the mat. The following are a few suggestions. Focus on the shelves rather than the border to begin with. Each shelf could be discussed separately and the children encouraged to look for particular details:

'How many teddies are wearing hats?'

'How many teddies are playing an instrument?'

'How many balls can you find?'

'How many teddies are playing some sort of sport?'

'What number is on the footballer's shirt?'

▲ Develop work on above and below. Ask questions such as:

'Which teddy is above . . . the crying baby teddy/footballer/trumpet player?'

● **Develop work on left and right.** Ask questions such as:

'Which teddy is on the left of . . . the trumpeter/clown/king/carol singer?'

Using the border

■ **Sorting and classifying**. The border gives good opportunities to sort and classify objects. Use a pile of assorted counters:

'Put a yellow counter on all the hats.'

'Put a green counter on all the shoes.'

'Put a red counter on all the balls.'

'Put a blue counter on all the instruments.'

'Put an orange counter on all the other clothes.'

'What is left? How could you sort these items?'

Using real teddies

Once the mat has been used to stimulate the children's interest, it is important to develop the work. Encourage children to bring in their own teddies to use for discussion and data gathering.

■ **Describing teddies.** Set out the bears in a row.

Describe one of them and invite the children to point out which bear you are talking about. Children soon get the idea and begin to describe the bears themselves.

▲ **Sorting teddies.** Make a collection of teddies in the classroom. Use some PE hoops to sort the teddies in as many ways as you can think of.

panda bears / not panda bears

brown bears / white bears

bears wearing clothes / bears without clothes

● **Ordering teddies.** Order the bears according to height or some other criterion.

■ **Measuring teddies.** Measure teddy bears and make them suitable items using any available materials. For example, a bed using a shoe box, a chair using multilink, card, wood, etc, an item of clothing: hat, scarf, coat, etc.

▲ **Telling stories about bears.** Make a collection of stories and story books about bears. The following are possibilities:

Caroline Bucknall: *One Bear in the Picture*, Macmillan

Caroline Bucknall: *One Bear All Alone*, Macmillan

Jill Murphy: *Whatever Next?*, Macmillan

Allan Ahlberg and Colin McNaughton: *Bear's Birthday*, Walker Books

R Maris: *Are you there, Bear?* Walker Books

Stan and Jan Berenstain: *Bears in the Night*, Harper Collins

Niki Yektai: *Bears in Paris,* Puffin

There are, of course, all the traditional stories and songs to be considered such as 'Goldilocks and the three bears' and 'The teddy bears' picnic'.

Support activities

There are numerous opportunities for matching and counting.

Extension activities

■ **Classifying and recording data**. When children are confident about sorting and classifying the objects in the border, repeat with the bears on the shelves. Put counters on the objects to be counted. The counters could then be transferred onto a pre-prepared chart to produce a graph of some of the items:

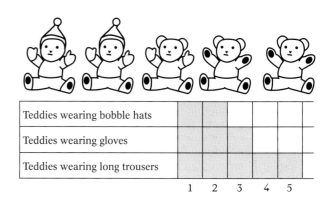

	1	2	3	4	5
Teddies wearing bobble hats					
Teddies wearing gloves					
Teddies wearing long trousers					

Teatime

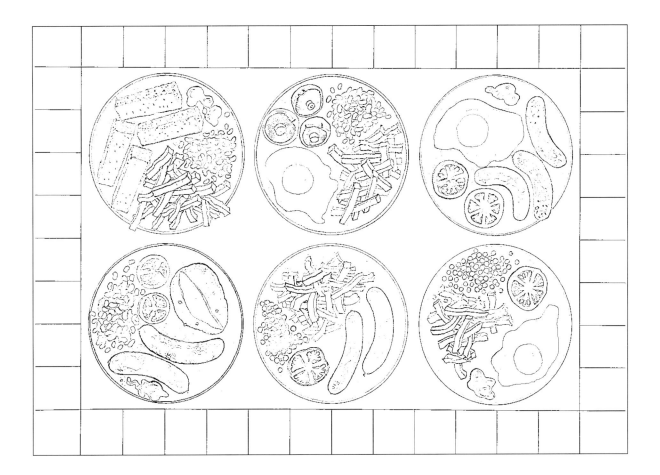

Mathematical concepts

Number bonds, counting, addition and subtraction, simple money, sorting, ordering, matching, data handling

Description of the mat

Within the centre section there are 6 plates, all the same size. On each plate is a selection of food. This food has been specifically chosen to be easily recognisable by shape and colour, but many children's favourite foods will be different according to their experience. It would be useful to produce actual plates of food to reflect their choices. These might include different items, such as rice, sweet potato, pasta and chick peas. These could be used instead of the mat illustrations for some activities, and the questions adapted accordingly.

The plates have coloured rims to help children to identify the plate when they discuss the food on each. The food items are eggs, sausages, tomatoes, tomato sauce, chips, peas, beans, mushrooms and fish fingers.

The tomatoes have been cut in half. See if the children comment, and if not you could introduce the idea that two halves make one tomato. Demonstrate. Notice also the different patterns on the tomato halves.

The border is a simple pattern of three colours repeated in order. It can be used for simple games and activities using counters and dice. Numbers or words could be written in.

Useful resources

Coloured cubes, paper plates, menus, plastic food (or felt or dough food), magazines with pictures of food, egg boxes, templates and shapes for drawing food items like eggs and sausages

Teacher's notes

Using the plates in the centre

■ **Preliminary discussion.** Work with a group of six children. Give each child, or let each make, a name card to place by the plate of their choice. They could take turns to choose.

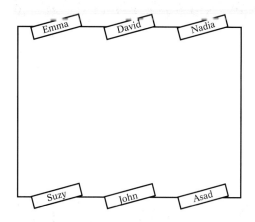

Start a general discussion where children must contribute according to the food on their plate:

'Who has beans?' 'Who has an egg?' 'Who has fish fingers?' 'Who does not have sausage?' 'Who does not have tomato sauce?'

'Who has chips and egg?' 'Who has 5 different items?'

'Who has something no one else has?' 'Who has the most tomato?' 'Who has the most sausages?'

Once the children are very familiar with the mat the following sorts of question could be asked.

'Who is sitting opposite a person with an egg?' 'Who is sitting opposite a person with chips?' 'Who is sitting next to (beside) a person with sausages?'

'How many sausages are there altogether?' 'How many pieces of tomato? . . . eggs? . . . mushrooms?'

The mat should be used several times for discussion and comparison. Introduce the vocabulary more than, less than, left, right, above and below. Turning the mat will result in changing the discussion and reinforcing the vocabulary used. No formal recording is necessary to begin with.

Using the border

■ **Matching and counting.** Ask children to match cubes to the colours around the border to form a repeating triple pattern. Sort and count the number of each colour.

▲ **Reconstructing the border pattern.** Cut out some pieces of card the appropriate size to cover blocks of the border, for example a rectangle to cover four neighbouring squares.

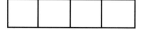

Use this to cover a section of the border. Ask children to tell you what is covered in the correct order.

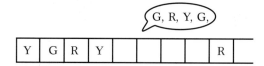

● **Fill your plate: a game for 2 to 4 players.** You need a standard 1–6 dice, counters or cubes or representative food for each of the three colours in the border (eg tomatoes, eggs and peas, about 12 of each kind – these could be made by the children in preparation for the game), a paper plate for each player, a different coloured counter for each player.

Each player is aiming to collect three of each item of food or coloured counter. Players put their counters at a chosen starting point. They take it in turns to roll the dice and move around the border collecting an appropriate item on each landing. The first player to collect three of each item is the winner.

- **Pattern play: a game for 2 to 4 players**. You need a standard 1–6 dice, a different coloured counter and a strip of squares for each player, identical in size with a section of the border, and yellow, red and green pens or pencils. Players aim to colour the squares of their strip in the correct order to match the border.

| G | R | | G | R | | G | R | |

Take turns to roll the dice and move a counter the appropriate number of squares around the border. You colour the next square of your strip if the colour needed matches the square you have landed on.

It may help to letter the squares as shown to help the children remember the pattern order. The length of the pattern you give the children should depend on their attention span.

Activities using pretend food

- **Discussion and counting**. Provide each child with a paper plate and invite them to draw a meal, or cut out pictures of foods, or use modelling material to make food to put on their plates. Once each child has made a plate, group discussion can take place as for the mat.

- ▲ **A classroom café.** Set up a café in the classroom. Price each item of food on the menu and introduce a simple money activity.

- ● **Developing addition and subtraction.** Develop addition and subtraction using pretend sausages or eggs. A 6- or 12-egg box with wooden or plastic eggs can be a really useful resource. (Cotton reels could be used if eggs are not available, or make some with modelling material.) Start with the box full, and play games pretending to eat or use some of the eggs.

'Here are 6 eggs. I eat 3, how many are left?'

'I did have 6 eggs. Now there are only 2. How many have been eaten?'

Tell stories about eggs, to introduce simple problem solving. 'I collected two eggs yesterday and two more today. How many eggs are there in my box? . . . Open it and see.'

- **Number patterns based on 'Ten fat sausages'.** Set up a pan with felt or model sausages. Sing 'Ten fat sausages sizzling in the pan'. Remove some and ask how many are left. Develop the idea of number patterns by first removing 2, then another 2, then another 2, and so on.

Then count down in 2s:

10, 8, 6, 4, 2, 0

- ▲ **Recording**. You could produce simple recording sheets for children to draw food items on plates or in pans. Two examples are shown below. Make sure that you leave enough space for children to draw in the missing items.

Seven silly sausages sizzling in the pan

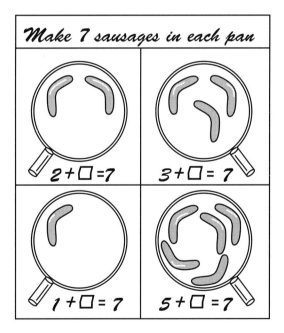

Cakes on a plate

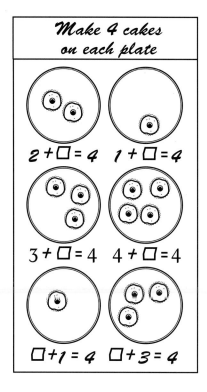

Make 4 cakes
on each plate

$2 + \square = 4$ $1 + \square = 4$

$3 + \square = 4$ $4 + \square = 4$

$\square + 1 = 4$ $\square + 3 = 4$

Extension activities

■ **Recording items on the plates.** Simple recording can be introduced to show the items on the plates. A copy of the square grid could be used for this.

Other collections of 4 or more items can be recorded on the sheet. For example, use it to record favourite meal choices of the children.

A graph to show how many people have these items

(bar chart with vertical scale 1–10; bars for: eggs = 3, mushrooms = 0, beans = 3, chips = 4)

A graph to show the total number of food items used

(bar chart with vertical scale 1–10; bars for: eggs = 3, saus = 7, mushrooms = 3, jacket potato = 1)

Snakes

Mathematical concepts

Length, number, pattern, left and right

Description of the mat

The centre of the mat contains four snakes, two facing left and two facing right. Each snake is coloured differently to highlight pattern and number. Around the centre are a selection of coloured rods which can be used to compare lengths.

The patterns on opposite sides of the border are intended to be decorative, but can also be used to provoke discussion. The sides are symmetrical. The four corners are marked with the numbers 1, 2, 3, 4.

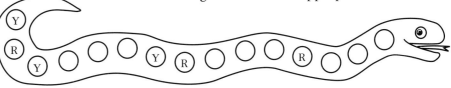

Useful resources

Plasticine or clay, counters, string, ribbon, sticks, Cuisenaire rods, copy of the mat

Teacher's notes

Using the snakes

■ **Discussing the snakes.** Start by talking about the mat. Count the snakes. Focus on each of the snakes individually and describe the patterns and count eyes.

Talk about left and right and which snakes face left.

▲ **Using counters to count spots and markings.** Use coloured counters to match the colours on the mat. Place appropriate counters onto spots and markings on the snakes, then count and compare.

● **Using snakes as number lines.** Use each of the snakes as a number line, marking the numbers using an ohp pen or chinograph pencil. The alternating pattern on the top snake gives an opportunity to discuss odd and even numbers.

Mark some of the numbers on the mat or a copy of it and ask children to fill in the missing numbers.

■ **Exploring patterns.** You can discuss the patterns in the markings on the snakes. Lay out some counters in a pattern formation on the blue snake and ask the children to add the missing counters in the appropriate colours.

▲ **Snake stories.** Make a collection of snake stories such as:

Nan Bodsworth: *A Nice Walk in the Jungle*, Picture Puffin

Rudyard Kipling: *The Jungle Book*

Using the coloured rods

■ **Matching and comparing lengths.** Produce a set of card, wood or paper sticks to match the lengths of the coloured rods. These could be in the correct colours, but a neutral colour would challenge the children to consider length rather than colour. Ask the children to match the lengths. The actual sticks could then be ordered and sorted and combined to produce lengths longer than, shorter than, the same length as , ...

Ask questions such as:

'Which stick is the longest?'

'Find two sticks which together are longer than the purple rod.'

A similar activity could be done using pieces of string or ribbon cut to length.

Snakes

NCM Module 1 Mats

▲ **Comparing lengths using Cuisenaire rods.**
Each Cuisenaire rod is an exact number of
centimetres long and is coloured according to
length. A chosen length can be made using
various combinations of rods.

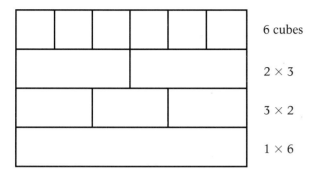

6 cubes

2 × 3

3 × 2

1 × 6

Making snakes

Snake making activities develop co-ordination and
estimation skills in children.

■ **Snakes from dough, clay or plasticine.** Give
children dough, clay or plasticine and ask them
to make snakes the same length as the four
snakes on the mat. The materials can be put
directly on the mat which can then be wiped
clean after use.

▲ **Making thick and thin snakes.** Ask the children
to make four snakes out of plasticine or a similar
material such that one is long and thin, one is
short and thin, one is long and fat and one is
short and fat.

● **Ordering plasticine snakes.** Ask children to
make four snakes of different lengths and to
order them.

■ **Making plasticine snakes of a given length**.
Give each child two pieces of string, one long
and one short, and ask them to make two snakes
exactly the same lengths as the pieces of string.

▲ **Making longer or shorter snakes.** Give each
child a piece of string and ask them to make two
snakes, one longer than the string and the other
shorter.

● **Making letters from snakes.** Use plasticine
snakes to make shapes of letters or simple
words.

■ **Making collage snakes.** Collage snakes can be
made by cutting and sticking pieces of fabric.

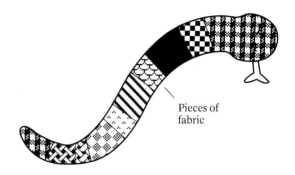

Pieces of
fabric

▲ **A number snake to use as a number line and
mobile.** Draw a spiral on a circle of paper and
decorate it. Cut it to make a long snake. This
type of snake can make a number line and
mobile.

● **Snakes from fabric**. Use old stockings or tights
and stuff them to make snakes. The end can be
folded in and sewn. Decorate with felt shapes
glued or sewn to make a pattern. You can use
beads for the eyes and felt for the tongue.

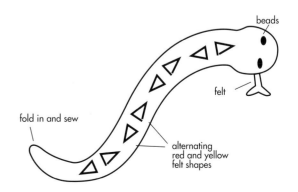

beads

felt

fold in and sew

alternating
red and yellow
felt shapes

- ■ **Printing patterns.** Print a pattern inside a pre-drawn snake outline using potato prints, sponge shapes, cotton reels or corks.

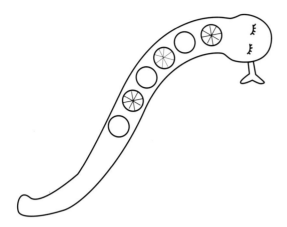

- ▲ **Patterned snakes from plasticine.** Roll out a thin snake using clay or plasticine. Flatten this slightly then press items into the clay to produce a pattern. Once this is dry use it to produce a rubbing by placing paper over the dry clay snake then rubbing with wax crayon or paint over the snake and print with it.

Extension activities

- ■ **Making snakes twice or half as long.** Give each child a piece of string and ask them to make a plasticine snake which is twice as long as the string or half the length of the string.

Sticks

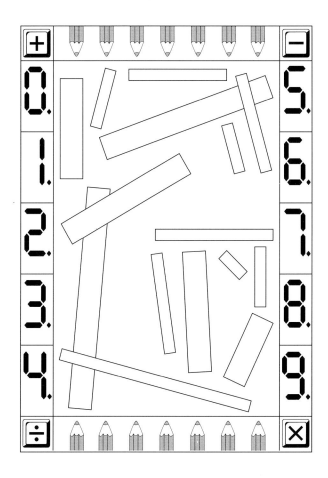

There are three sections to the border. The long sides depict the digits 0 to 9 as displayed in light bars on a calculator. It is important that children are exposed to a variety of ways of recording numbers, as well as the standard written numbers.

$$0\ 1\ 2\ 3\ 4\ 5\ 6\ 7\ 8\ 9$$

Each of the four corners displays one of the four operation keys.

The top and bottom sections each show a selection of seven coloured crayons in three colours.

Useful resources

Cuisenaire rods, Deines apparatus, straws and pipe cleaners, geostrips, lollipop sticks, card strips in various colours – a set to match the mat would be very useful, calculators

Teacher's notes

Using the sticks

- ■ **Counting the sticks**. Count the sticks according to colour, thickness, etc.

- ▲ **Discussing comparative lengths and widths**. Discuss comparisons such as these:

  'Which stick is the longest/shortest?'

  'Choose a stick. How many are longer/shorter than your chosen stick?'

  'Are any of the sticks the same length?'

  'Which sticks are wider than this stick?'

- ● **Using straws or card to compare lengths**. Cut art straws or strips of card to match the length of the sticks. Compare the lengths of sticks.

- ■ **Grouping the sticks**. Use counters or strips of card or a marker pen to mark the sticks in groups.

Mathematical concepts

Number, length and comparison, matching, sorting, calculator awareness and familiarisation, shape, pattern

Description of the mat

The centre of this mat has been designed to focus on length and the use of sticks for various practical activities. It shows a selection of 15 coloured sticks in two widths – thick (24 mm) and thin (12 mm) – and several lengths from 3 cm to 22 cm. There are:

Colour	Thick	Thin
orange	7 cm	10 cm, 10 cm
yellow	14 cm	10 cm, 5 cm
green	10 cm	20 cm, 6 cm
red	22 cm	12 cm
blue	12 cm	6 cm
purple	18 cm	3 cm

Mark all the thick/thin/yellow/... sticks.

Mark sticks shorter than this:

7cm

for example.

Use actual sticks, rods, straws, so that the child can make comparisons.

▲ **Comparing lengths using strips of paper**. Cut strips of paper to match each stick.

Make a staircase with the strips by drawing a line to build from. Start with only about five sticks, but gradually increase the number to be compared.

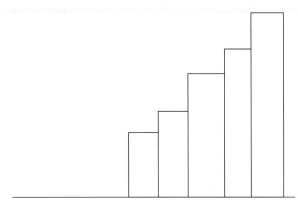

Use the strips to produce a stick picture.

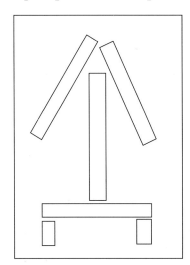

Cut out some sticks of your own to sort for thickness/length/colour.

Cut out pairs of identical sticks or strips of paper. Ask the children to match up the identical pairs.

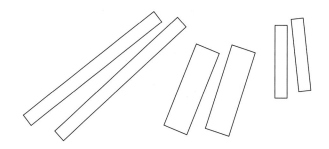

Collect pieces of ribbon to work with in the same way.

Make a collection of straight objects for comparison: pens, pencils, garden sticks, rulers, off-cuts of beading. Sort and order according to length and thickness.

The pencils in the border

■ **Counting using objects**. Count the pencils by covering each with a counter. Use matching coloured counters to count the number of each colour.

▲ **Comparing numbers**. Compare the numbers of each colour of crayon at each end. Which end has most red/blue/yellow?

● **Exploring the number**. Discuss the number. Explore other ways of having seven coloured pencils. Begin by making seven using just two colours, then move on to three colours.

Collect sets of seven similar objects; sort and order them.

■ **Comparing lengths of pencils**. Use sets of seven pencils of different lengths. Order them according to length by matching the tips to the tips of the border crayons.

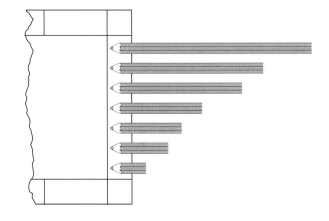

- ▲ **Introducing addition and subtraction**. Use coloured pencils to play games where you introduce addition and subtraction. Use colour to help children recognise the changes. For example, start with 5 red pencils and add 2 blue ones. How many are there now?

- ● **Making a pattern**. Introduce a simple pattern by laying out crayons in patterns to be continued by the children.

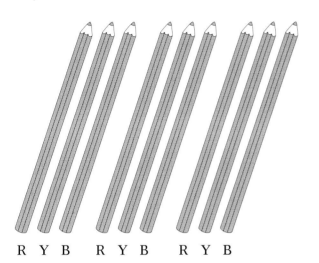

R Y B   R Y B   R Y B

Using the calculator numbers

You will need short lengths of stick, modelling match sticks or pieces of drinking straw to use as light bars. Lollipop sticks may be used but they take up a lot of room.

Each child will need access to a calculator. Recording of calculator digits can be done using resource sheet 6 in the Teacher's resource book.

- ■ **Matching standard digits and calculator digits**. Use the 0 to 9 cards of the number cards with the games. Ask children to match them to the digits on the mat by covering or laying them side by side.

- ▲ **Using a calculator**. Challenge the children to produce the digits on a calculator display panel. Ask questions such as the following according to the development of the child.

'Can you make a row of 8s appear in the display?'

'How many will fit on?'

'Can you change these to 5s?'

'Can you display this pattern of numbers: 121212 . . . or 345345 . . . ? What will come next? How about 12345678? Try some more.'

'What number displays can you make using only 0 and 1? Here are some to start you off:

01010101   10101010   00110011 11101110'

'Make a number display then ask your friend to copy it.'

- ● **Making calculator digits from straws**. Prepare some straws which children can use as light bars. Ask them to make a set of digits similar to those on the calculator. You will need glue, card, scissors and straws.

'How many pieces do you need to make each number?'

'Which digit uses most? . . . least? How many does 5 use? Which other number uses 5? Find ways of recording the numbers.'

You could roll a slab of plasticine and press sticks into it, or stick strands of plasticine over each of the numbers on the mat.

Using the operation keys

- ■ **Exploring the operation keys on a calculator**. Ask children to explore the four operation keys and to try to find out what happens when you use these with the number keys.

Discuss the importance of the = key and introduce its use. Use plasticine to reproduce the signs and symbols found on the calculator. These could also be pressed into slabs of plasticine.

Support activities

- ■ **Comparing lengths**. Give the child a Cuisenaire rod or Deines apparatus and ask them to find a stick on the mat which is longer/shorter/the same length as theirs.

Sweet shop

Mathematical concepts

Number, money, shape, sorting, number bonds, pattern, size, introducing addition, subtraction, multiplication and division

Description of the mat

There are six jars each containing a different sort of sweet of different shapes and colours. The jars are of different heights.

The border is divided into four sections each displaying a variety of coins with values up to 10p. The corners display 1p, 2p, 5p and 10p coins. The coins can be used for matching and collecting into piles on the corners.

The other side of the mat is identical in layout, but the jars and the border are all empty to allow for detailed practical activities.

Useful resources

A collection of coins, preferably real, with sufficient of each to match the mat, some empty jars and a collection of sweets to put in them, items to represent sweets for sorting, counting and weighing, stencils, sweet jars resource sheet

Teacher's notes

Using the jars

- **Introductory discussion**. Start with a discussion of the contents of each jar. Ask the children to talk about the colours of the sweets, their shapes and the sizes of the jars. Ask questions such as:

  'How many jars contain some yellow sweets?'

  'Which jar contains some red sweets?'

  'Which jar is the tallest?'

  'Describe what is in the first jar.'

Counting and comparisons can be made by using cubes or counters for the number in each jar. Discuss which jar has more than or less than another. The blank boxes below each jar can be used to write in numbers using an ohp pen.

Using the border

- **Matching coins**. Match coins to the money around the border. Count how many there are of each type. Look at a section at a time.

  Introduce simple exchange of coins, for example, 5 × 1p coins for 1 × 5p.

- ▲ **A simple investigation**. Ask children how many ways they can make 5p or 10p. Help them find appropriate ways to record their ideas.

- ● **Sorting and displaying the coins**. Match coins to the money in the border. Ask the children to sort them. Use the Square grid mat to record how many coins there are of each value.

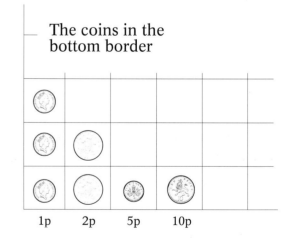

The coins in the bottom border

| 1p | 2p | 5p | 10p |

Using the empty side

This side provides somewhere to sort sweets and create your own collections.

- ■ **Sorting sweets**. Give the children a collection of sweets with a maximum of six different types. Ask them to sort these into jars. Use actual jars or the ones on the mat. Give them rules and ask questions, for example,

  'Put six sweets in each jar.'

  'Put the red sweets in the first jar.'

  'Fill each jar with a different sweet.'

  'How many sweets are there in each jar?'

  'Which jar has the most sweets?'

- ▲ **Introducing addition and subtraction**. Put a random selection of sweets in each jar. Ask the children to add extra sweets or take some away so that there are, say, exactly eight sweets in each jar.

- ● **Introducing simple multiplication**. Develop the idea of multiplication by asking questions such as:

  'If there are two candy sticks in each jar, how many will there be altogether?'

  Encourage children to try it out using sticks or similar items.

- ■ **Recording**. Children can use copies of the resource sheet to draw their own sweets in jars using a stencil and to record the number in the box underneath.

- ▲ **Developing number bonds**. Set simple problem solving tasks using the six jars on the mat or a collection of actual jars and sweets, for example,

  'There are 3 jars and I have 12 candy sticks. How can I share them out?'

  This could develop into an exciting investigation for children and provide opportunities to introduce simple division and the vocabulary of sharing. It is important that you choose appropriate numbers to enable division to take place.

The children can record their results on the resource sheet.

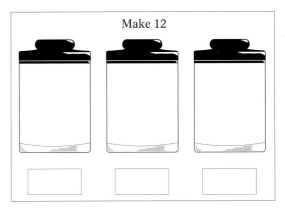

Make 12

- ● **Producing a collage or display**. Produce a sweet shop collage or display where each child is asked to cut out a sugar paper shape of a jar and put a selection of sweets inside. Ask them to talk about their sweet jar and compare it with jars made by other children. Restrict the materials available so that jars have similarity and difference. Stencils can be used to draw sweets.

  You can encourage children to classify the jars according to various criteria such as colour, number of sweets and shape.

Extension activities

- ■ **Exploring subtraction**. Begin to explore subtraction in a structured way. Invite the children to put six sweets in each of the six jars. Tell them to sell or eat sweets from the jars and count how many are left. Record what has happened below each jar, for example,

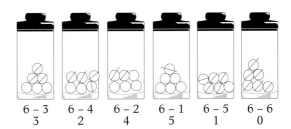

6 − 3   6 − 4   6 − 2   6 − 1   6 − 5   6 − 6
3   2   4   5   1   0

Now ask lots of questions such as:

'Which jar has the least sweets?'

Alternatively start with a different number in each jar and take away an identical number from each jar.

▲ **Problem solving using subtraction**. Put at least three sweets in each jar.

6 – 3 = 3   8 – 5 = 3   3 – 0 = 3   4 – 1 = 3   9 – 6 = 3   7 – 4 = 3

Tell the children to take away sweets from the jars so that there are three sweets left in each.

● **Making 15p**. Ask the children to cover all the money in the border with the appropriate coins. Ask a child to pick up 15p. There are several ways in which this can be done. Repeat the process. How many times can you pick up 15p? How much is left at the end? You can also do this activity with other totals.

■ **Making 20p**. Match coins to the money in the border. Ask which sections it is possible to take 20p from. Ask how much is left in each section.

▲ **Buying sweets**. Once children are able to relate to coins, you could give the sweets in the jars a price as a basis for interactive play.

2p   4p   3p   5p   1p   2p

Various investigations can then be explored such as:

'You have 10p to spend. What can you buy?'

'How many 4p candy bars can you buy with 10p?'

● **Problem solving with coins**. Put a set of coins in each section of the border on the reverse of the mat as shown. Say that each section should contain 10p, but there is a coin missing from each one. Ask the children to identify the missing coins.

	5p	2p	1p	
2p				5p
2p				
1p				
	5p	2p	2p	

Name..................

Sweet Shop

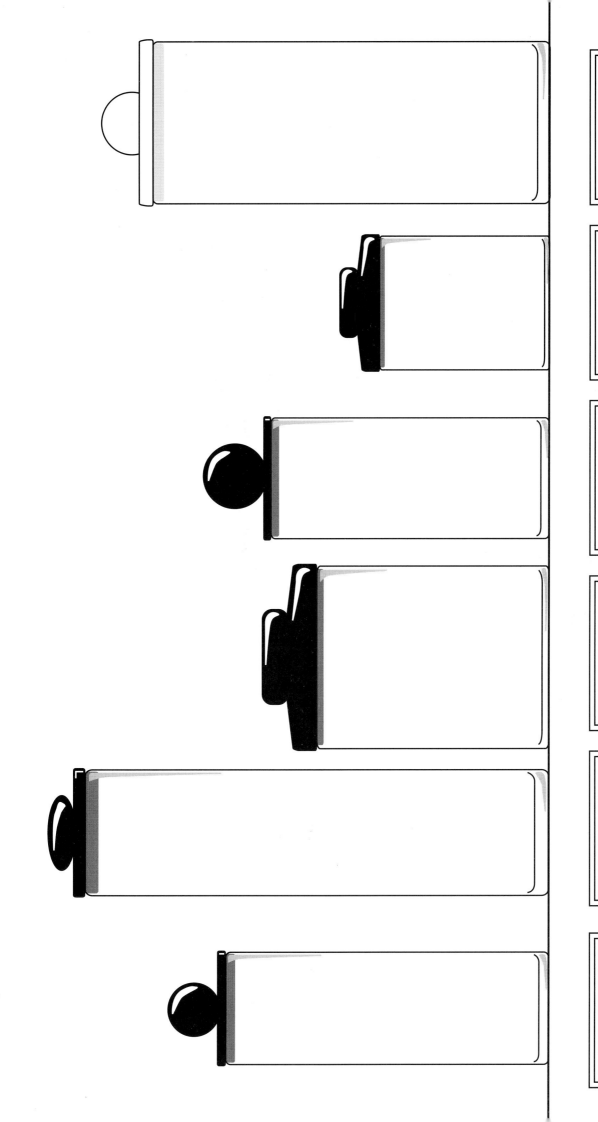

NCM Module 1 Mats

Jewellery

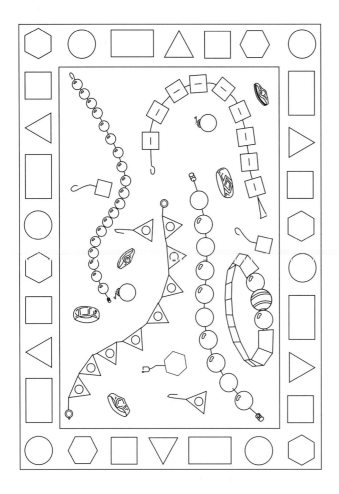

Mathematical concepts

Number, pattern, length, shape, matching, sorting

Description of the mat

There are four necklaces, two with round beads – one large and one small, a triangular design and a square design. Each has a different number of beads. The earrings are designed to introduce pairs and counting in twos.

The colouring of the necklaces is designed to stimulate discussion about pattern. There is a simple alternating pattern on the small beads, a reversing pattern on the triangular necklace.

The border contains a repeated pattern of rectangles, equilateral triangles, squares, hexagons and circles. They are coloured red, yellow and blue so that there are two of each shape in each colour for matching in pairs.

Useful resources

A collection of play jewellery – necklaces, bracelets, rings and earrings, belts, laces, beads of various shapes, lengths and sizes (including cubes) or multilink, counters, stencils, a standard dice, the jewellery resource sheets

Teacher's notes

- **Introductory discussion.** A general discussion should take place with the children to encourage observation. There is potential for counting necklaces, rings, earrings and bracelets, and identifying items by describing shapes and colours. Ask questions such as:

  'Point to the necklace with square faced beads. Find some square faced beads in the bead box.'

  'Which earrings would you wear with the triangular shaped necklace?'

  'How many beads are there on the longest necklace?'

  'How many earrings can you find? Match the pairs.'

- ▲ **Exploring patterns.** Encourage discussion of the patterns in the colours of the beads of the necklaces. The white necklace can be used for creating a child's own design using coloured counters.

Using the border

- **Using counters to help with counting.** Cover each shape with a counter of the matching colour. Count how many there are of each colour.

Cover all triangles with a red counter. How many are there?

Cover all shapes with four sides with a green counter. How many are there?

How many yellow hexagons are there?

▲ **Investigating patterns.** Put a multilink cube or cube bead on each of the squares around the border, matching the colours. Then make a necklace or tower with these cubes.

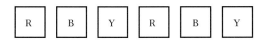

Discuss what is meant by a pattern. Are these both patterns?

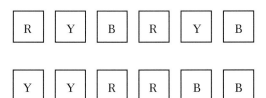

Ask children to make as many different patterns as they can. Children can record their patterns on the six beads resource sheet.

Ask how the patterns will continue if the necklaces are twice as long.

You could also begin to introduce the idea of a symmetrical pattern such as,

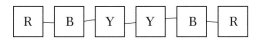

It is often helpful to restrict the type of pattern that is allowed with very young children as this helps them to focus on a particular aspect of it.

● **Collecting beads: a game for 2 to 4 players.** You need multilink cubes in red, yellow and blue or beads of the same shapes or colours as the shapes in the border, a standard 1–6 dice (or a dice labelled 1, 2, 3, 1, 2, 3, if necessary to start with) and a different coloured counter for each player.

Players all put their counters on one of the corners. Take it in turns to roll the dice and move your counter the appropriate number of shapes around the board. Each time you land on a shape, collect a cube or bead of the matching colour or shape from the collection. Once the agreed number has been collected, say 10, then the child threads the beads or cubes onto string or lace to produce a necklace. Pattern should be encouraged where possible.

An alternative would be to have the children collect six objects of each colour or shape. Vary this according to the children's ability and level of concentration.

■ **Design and make a necklace: a game for 2 to 4 players.** You need a lace each and multilink cubes in red, yellow and blue or beads of the same shapes or colours as the shapes in the border, a standard 1–6 dice and a different coloured counter each.

Each take a lace and start a necklace using four beads in two colours or two shapes. Describe your pattern to the other children in the group. Each put your counter in one of the corners. Take it in turns to roll the dice and move that number of shapes around the board. You may choose whether to move backwards or forwards each turn. The aim is to collect shapes or colours to continue the necklace you are making. You may only add a bead if it matches the pattern you are making.

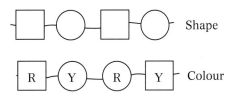

The winner is the first player to have a necklace 9 beads long, say.

Making jewellery

Children can use a collection of necklaces to make comparisons of length, introducing longer than and shorter than and ordering several items according to length.

■ **Making necklaces using laces and beads.** Help the children make a necklace using laces and 8 or 10 beads. Then ask them to make one using just two colours.

If you have access to enough beads the children could investigate ways of making necklaces using the same two colours but in a variety of combinations. The necklaces resource sheet has nine beads, but you may want to vary the number to suit the ability of the children and the resources you have available. Multilink cubes are an ideal resource if beads are unavailable.

▲ **Drawing or printing necklaces.** Use a stencil or beads to design and draw necklaces. Printing with corks or cotton reels or other round objects can be great fun.

● **Making necklaces from pasta or straws**. Pasta or straws can be painted in different colours and threaded onto string or laces to help reinforce the idea of pattern building.

■ **Making beads from paper.** Roll paper around pencils and glue it to produce 'beads' of different lengths. These can be used to make necklaces. You could also use polystyrene chips and other waste items such as cardboard discs.

Support activities

■ **Reinforcing number.** If you need to reinforce number, the necklace beads may be used as number lines and numbers inserted by the teacher or child for the purpose of counting.

Extension activities

■ **Number patterns.** You can ask children to insert the missing numbers in number patterns given on the ten beads resource sheet, such as:

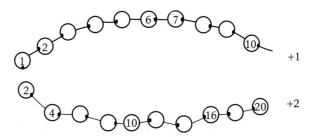

Children can use a number line to help with these.

Six beads

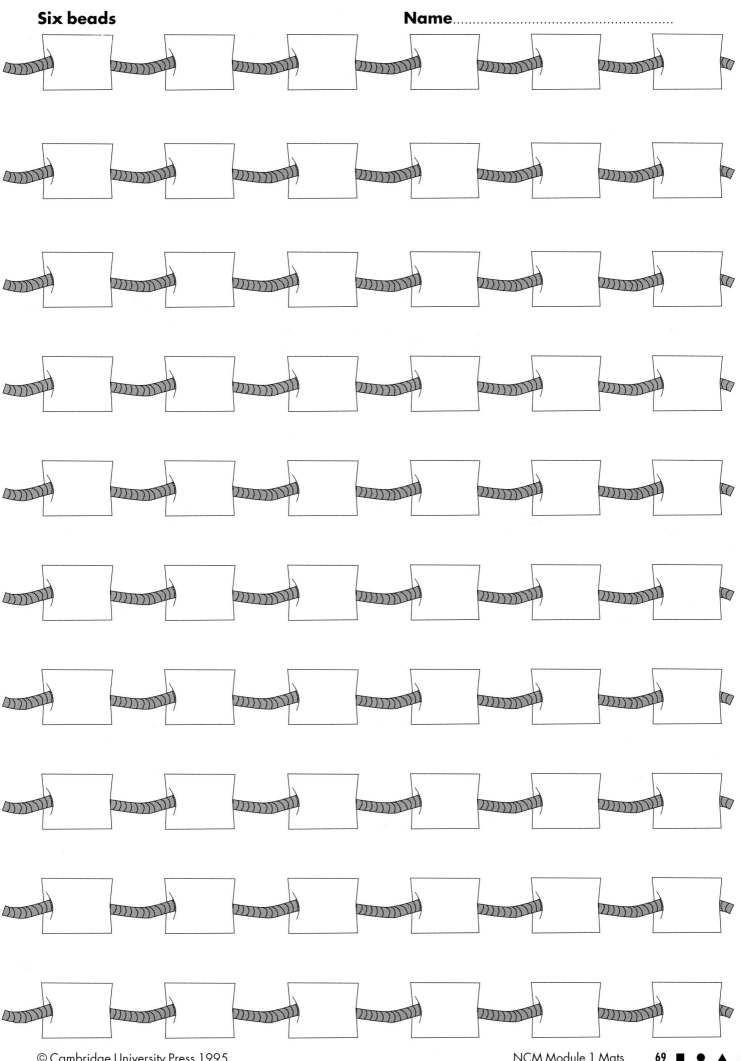

Ten beads   **Name**...

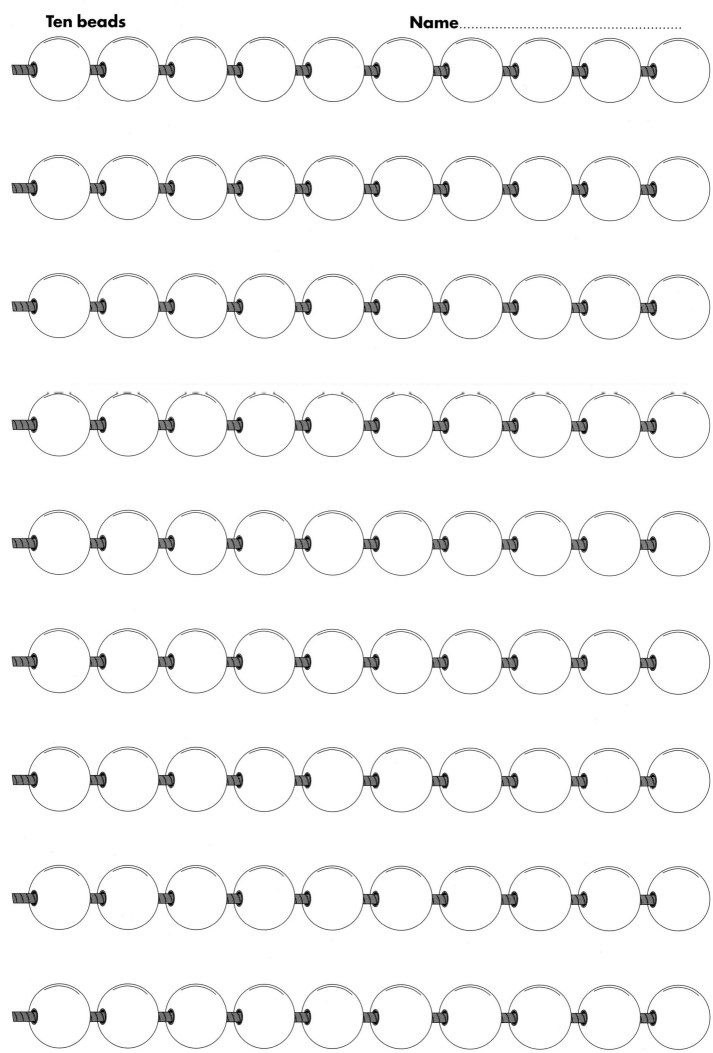

Shapes

Mathematical concepts

Shape, number, pattern, size and colour, collecting, organising and recording data

Description of the mat

This mat has been designed to provide a focus for discussion and practical activities relating to the four basic shapes, triangle, square, rectangle, hexagon.

Three of the shapes have exact measurements in centimetres. This means that children could use centicubes for measuring the lengths of sides and perimeters.

triangle	sides 12 cm
square	sides 10.5 cm
rectangle	sides 15 cm, 8 cm
hexagon	sides 6 cm

The border has several copies of each of the four shapes. These are drawn in two sizes, large and small, to give further opportunity for sorting and counting. Each corner contains a version of one of the four shapes in white on a purple background. The numbers of shapes illustrated on the sides, excluding the corners are:

	triangle	square	rectangle	hexagon
Large	6	7	2	3
Small	4	5	4	5
Total	10	12	6	8

Each section of the border contains different selections to allow for comparison.

Top	10 shapes	3 triangles, 4 squares, 1 rectangle, 2 hexagons
Bottom	10 shapes	3 triangles, 1 square, 4 rectangles, 2 hexagons
Left	8 shapes	2 triangles, 4 squares, 0 rectangles, 2 hexagons
Right	8 shapes	2 triangles, 3 squares, 1 rectangle, 2 hexagons

Useful resources

Counters, stencils, a box of shapes including the four basic shapes here in a variety of sizes and colours, stencils of the four shapes, shapes resource sheet and copy of the mat

Teacher's notes

■ **Discussion of the shapes.** Explore the mat in general discussion. Count the sides, and angles of the shapes. Ask children to draw round each shape with their fingers. Ask them to point to various shapes, for example a large square or a small hexagon.

▲ **'Can you find' games.** 'Can you find' games are fun to play with a group, for example,

'Can you find a large square next to a large triangle?' [1 way]

'Can you find a triangle between a square and a rectangle?' [3 ways]

'Can you find three large shapes next to each other?' [1 way]

The children should be encouraged to describe a selection of shapes for the rest of the group to find.

● **Using counters to cover shapes and count them**. Each child collects counters to put inside each of the centre shapes, for example 6 blue counters in the triangle, and 10 red counters in the rectangle. Ask the children to use the counters to cover particular shapes, for example cover the large triangles in the border using the blue counters. Ask questions such as:

'Do you have enough?' [yes]

'How many more blue counters do you need to cover the small triangles?' [4] 'How many border shapes are now covered?' [10]

'What shape have you covered?' [triangle]

Name ....

Shapes

Using the border

■ **A game for 2 to 4 players**. Each player has a different coloured counter which is put on one of the corners. A dice or spinner is needed, marked with the four shapes (and two blanks if a six-sided dice is used).

Players take it in turns to roll the dice and to move around the border to the next shape which matches the one which was rolled. For example, if Ivan starts on the rectangle and rolls a hexagon he moves three places to the small hexagon.

Play continues until a player gets all the way around the border and lands back on his or her starting shape. For example, Ivan needs a rectangle to finish.

Group activities with moderate supervision

■ **Making pictures using shapes.** Children can make pictures using the four shapes in a variety of ways. They can use the stencil to draw the shapes, then colour or paint them. Alternatively they could use sticky paper cut into the shapes.

Flower from square and 4 triangles

▲ **Making patterns using shapes.** Provide a set of sponge shapes and trays of paint for pattern printing. Fluorescent paint on black paper is always very effective. Sample designs may be provided or demonstrated.

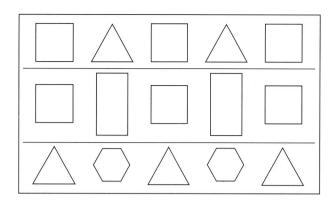

The shapes resource sheet

■ **Counting the number of shapes using cubes.** Suggest that children use coloured stacking cubes (for example, multilink or unifix). Cover each border shape with an appropriate coloured cube (blue for triangle, yellow for square, red for rectangle, green for hexagon) and then build a tower of each colour on the corresponding corner shape (a tower of 10 blue cubes on the corner triangle, etc.).

Pupils could then make a data chart showing the number of each colour using squared paper or a copy of the shapes resource sheet. The blank graph can be used for recording the numbers of shapes on a single side of the border, or the whole border. This could lead to discussion using more/less than, most, least.

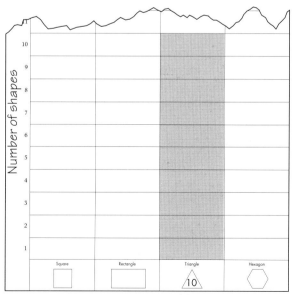

blue

Shapes

	Square	Rectangle	Triangle	Hexagon
15				
14				
13				
12				
11				
10				
9				
8				
7				
6				
5				
4				
3				
2				
1				

■ ● ▲ ■ **74** NCM Module 1 Mats

Support activities

- **Sorting real shapes**. Give the children a bag of shapes and ask them to sort them so that the triangles are all placed on the triangle on the mat, etc.

- ▲ **Colouring shapes on a copy of the mat.** Children could colour all the squares in the border yellow and similarly for the other shapes.

Extension activities

- **Exploring numbers using shapes**. Provide each child with a particular shape or choice of shape, then write numbers in each corner. They have to write the total in the middle. Any numbers could be inserted depending on what you wish to focus on or the ability of the child. Provide cubes or counters to place on each corner for children to build appropriate towers for totalling. Obviously, the more corners the shape has, the more challenging the puzzle.

This can be developed by asking children to make each shape total a set number, for example. 12, 18, ... Children should be encouraged to use a calculator.

Total 18

If you are lucky you may get:

$4\frac{1}{2}$	$4\frac{1}{2}$
$4\frac{1}{2}$	$4\frac{1}{2}$

You can also set challenges by asking children to use:

only even numbers, odd numbers, numbers less than 10, . . .

numbers greater than 5 but less than 10,

other groups of numbers, for example consecutive numbers for more able children, the same number on each corner.

- ▲ **Making and using bar charts.** Prepare some bags or boxes each with a maximum of 12 shapes in it. Alternatively ask the children to do this. Using the resource sheet each child in the group could make a chart showing different shapes in one of the bags or boxes.

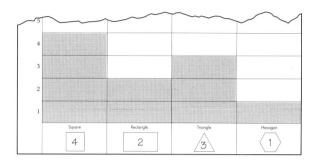

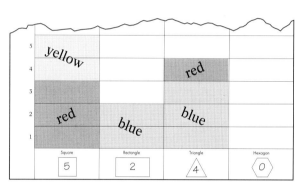

Once the graphs have been correctly drawn they can be displayed with the bags or boxes and the children can then match the data to the correct bag.